The Hong Kong Economic Policy Studies Series

PRIVATIZI

AND

SEWAGE SERVICES

PRIVATIZING WATER
AND
SEWAGE SERVICES

Pun-Lee Lam
Yue-Cheong Chan

Published for
The Hong Kong Centre for Economic Research
The Hong Kong Economic Policy Studies Forum
by

City University of Hong Kong Press

First published 1997
Printed in Hong Kong

ISBN 962-937-010-7

Published by
City University of Hong Kong Press
City University of Hong Kong
Tat Chee Avenue, Kowloon, Hong Kong

Internet: http://www.cityu.edu.hk/upress/
E-mail: upress@cityu.edu.hk

The free-style calligraphy on the cover, *shui,* means "water" in Chinese.

Contents

Detailed Chapter Contents

Foreword

The key to the economic success of Hong Kong has been a business and policy environment that is simple, predictable and transparent. Experience shows that prosperity results from policies that protect private property rights, maintain open and competitive markets, and limit the role of the government.

The rapid structural change of Hong Kong's economy in recent years has generated considerable debate over the proper role of economic policy in the future. The restoration of sovereignty over Hong Kong from Britain to China has further complicated the debate. Anxiety persists as to whether the pre-1997 business and policy environment of Hong Kong will continue.

During this period of economic and political transition in Hong Kong, various interested parties will be re-assessing Hong Kong's existing economic policies. Inevitably, they will advocate an agenda aimed at altering the present policy-making framework to reshape the future course of public policy.

For this reason, it is of paramount importance for those familiar with economic affairs to reiterate the reasons behind the success of the economic system in the past, to identify what the challenges are for the future, to analyze and understand the economy sector by sector, and to develop appropriate policy solutions to achieve continued prosperity.

In a conversation with my colleague Y. F. Luk, we came upon the idea of inviting economists from universities in Hong Kong to take up the challenge of examining systematically the economic policy issues of Hong Kong. An expanding group of economists (The Hong Kong Economic Policy Studies Forum) met several times to give form and shape to our initial ideas. The Hong Kong Economic Policy Studies Project was then launched in 1996 with some 30 economists from the universities in Hong Kong and a few from overseas. This is the first time in Hong Kong history that a concerted public effort has been undertaken by academic economists in the territory. It represents a joint expression of our collective concerns, our hopes for a better Hong Kong, and our faith in the economic future.

The Hong Kong Centre for Economic Research is privileged to be co-ordinating this project. The unfailing support of many distinguished citizens in our endeavour and their words of encouragement are especially gratifying. We also thank the directors and editors of the City University of Hong Kong Press and The Commercial Press (H.K.) Ltd. for their enthusiasm and dedication which extends far beyond the call of duty.

Yue-Chim Richard Wong
Director
The Hong Kong Centre
for Economic Research

Foreword by the Series Editor

Water supply in Hong Kong has long been provided by the government. Hong Kong citizens on the average spend only a scanty percentage of their total expenditures on the use of water. It is probably because the amount is so small that people do not usually pay much attention to the industry of water services. They are not aware of whether its industrial structure is well organized, or whether its economic efficiency is agreeable. Most of them simply think government-run water supply services is a matter of course.

On the demand side, in the early post-war years people in Hong Kong experienced hard times of water shortage. They would be delighted only to find no restriction in the supply of water. However, as the Hong Kong economy develops rapidly, people are now more concerned with their personal health as well as environmental protection. Their demand for water services shifts from quantity to quality, and they pay much more attention to sewage treatment and waste disposal.

Water and sewerage systems are public utilities, and as such are similar to others such as electricity, gas, transportation, and telecommunications. All these industries require large sums of initial fixed capital investment, and enjoy the subsequent economies of scale. The only difference in Hong Kong is that these other industries are all privately owned and managed, subject to different modes of government regulation.

Do the supply of and the demand for water and sewage services match in Hong Kong? Currently water and sewage services are supplied separately by the Water Supplies Department and the Drainage Services Department. Are these government departments fulfilling the demand of the majority of the consumers efficiently? How do they compare with other suppliers of public utilities? Is there any room for improvement? These are some of the important issues discussed in this book.

The authors, Dr. Pun-lee Lam and Mr. Yue-cheong Chan, approach the issues from the standpoint of industrial economics, and provide a comprehensive analysis of water and sewage services in Hong Kong. They discover that in terms of prices, profits, and productivity, the current water supply industry is inferior to other public utilities. Likewise, consumers do not seem to find sewage services satisfactory either.

The authors concretely propose to sell the publicly owned water utilities and privatize water supply. As for sewage services, since the issue of externalities is more prominent, the authors propose contracting out the sewerage systems. They also discuss how to regulate the water industry after privatization, making references to the experience of water privatization in the U.K. and to the regulation of other public utilities in Hong Kong as well.

Water and sewage services affect all households and business sectors; they also affect the ecology of the territory. This book provides a new perspective on dealing with the relevant issues in terms of economic efficiency. It is a good reference not only for water and sewerage services, but also for the regulation of public utilities in general.

Y. F. Luk
School of Economics and Finance
The University of Hong Kong

Preface

We would like to express our gratitude to two anonymous referees for their generous and enlightening comments and advice. Their suggestions have greatly improved the structure of this study and expanded our knowledge of the field.

Throughout our study, we received kind support from many academics and legislators. They have provided us with a great deal of useful information. Finally, we would like to thank Karen Chan and Josephine Wu who have helped us obtaining data and documents about the water and sewerage industries in Hong Kong.

Pun-Lee Lam
Department of Business Studies
Hong Kong Polytechnic University

Yue-Cheong Chan
Department of Business Studies
Hong Kong Polytechnic University

List of Illustrations

Figures

Tables

Acronyms and Abbreviations

CHAPTER 1

Introduction

Water Supply and Sewerage Industries in Hong Kong

In Hong Kong two different government departments, the Water Supplies Department (WSD) and the Drainage Services Department, are responsible for water and sewage services.[1] Another department, the Environmental Protection Department (EPD), is responsible for monitoring the water quality in Hong Kong. With the exception of the water supply and sewage service, all public utilities in Hong Kong are owned and operated by private companies. After Hong Kong became a colony in 1841, it was initially assumed that water, like all other utilities, would be supplied by private enterprises. Since the financial prospects of such a private venture were not promising, the idea was not adopted by the representatives of the business community in the Executive Council, and the government had to borrow money from banks to build the territory's first water supply system in the 1850s. Hence, Hong Kong's water supply has been in public hands for more than 140 years.

Over the past two decades, the people of Hong Kong have become more aware of water pollution problems. The government now monitors the water quality in Victoria Harbour and Tolo Harbour. In 1989 the Drainage Services Department was set up and charged with the task of coming up with an efficient approach to resolving sewage and flooding problems in Hong Kong. It is also responsible for the design, construction, operation and maintenance of the sewerage, sewage treatment and disposal, and storm water drainage systems in the territory.

After years of discussion, a trading fund was introduced in 1994 in the Drainage Services Department for the provision of sewage services. The fund is supposed to help recover the cost of operating and maintaining the public sewerage systems. This trading fund was established primarily to facilitate the "polluter pays" charging scheme. As all the sewage charges collected are paid into the trading fund to recover the operation and maintenance costs of sewage services, the public can see how much it costs to collect, treat and dispose of sewage. The public would then understand how the level of charges is determined. In addition, the trading fund arrangement provides an alternative means to getting on with work to upgrade the sewerage systems as fast as possible. The general objective of the fund is to increase the transparency of operational performance within the Department and to allow more flexibility in matching its service levels against consumer demand. The government proposed that the trading fund should break even in five years, by the year 2000.

Performance of the Water Supply and Sewerage Industries

In terms of price, returns and productivity, the water supply industry in Hong Kong compares unfavourably with private utilities in Hong Kong. The financial performance of the Water Supplies Department has been deteriorating in recent years. The Department has not met its target return since 1984–85. With decreasing water sales and continuously rising staff and water purchase costs, water prices have had to be increased at a rate higher than the level of inflation. The government has also increased the subsidy provided for water supply. Despite the price increases, there are signs indicating that labour productivity, water quality and supply reliability are deteriorating. The problems of water leakage and over-supply have become more serious than before. Increases in capital investment are therefore required to improve the water supply systems.

The public has also been dissatisfied with the way sewage services are charged. Industrialists have said that sewage charges will "destroy industries" in Hong Kong. Because of over-estimation of the revenue from sewage charges, the trading fund is expected to be in huge deficit after its first year of operation. Significant increases in sewage charges would be required for the fund to break even in the future.

In early 1996 the Water Supplies Department proposed a 9% rise in water charges beginning in July. The Drainage Services Department also proposed a 15% increase in sewage charges and trade effluent surcharges beginning in August 1996. In addition, the Drainage Services Department estimated that there would be higher charges in order for the trading fund to break even by 2000, by 165% between 1997 and 2000. In fact, the sewage charging system based on operating costs was approved by the Legislative Council a few years ago. The proposed increases in water and sewage charges again aroused strong public resentment. On 27 June 1996, Legislative Council members passed a motion to block the proposed increase in water charges. A week later, they also blocked an increase in sewage charges, reversing their earlier endorsement. Government officials warned that the price freeze would mean a cut in water and sewage services provided to the public. The Sewage Service Trading Fund would continue to face a deficit and would fail to strike a balance by 2000. The officials argued that if the government had to provide additional financial resources from general revenue, it would violate the "polluter pays" principle endorsed by legislators a few years ago.

Scope

The increases in water and sewage charges proposed in 1996 rekindled the debate over the privatization of government enterprises. It has been argued that since the Water Supplies Department and the Drainage Services Department are both government enterprises which can neither fire unproductive employees nor deal with wage flexibility (*Ming Pao Daily News,* 22 May 1996), it is difficult to measure the efficiency or cost-effectiveness of these government departments. As consumers do not have the choice of using any other service providers, it seems unjustifiable to pass all the cost increases on to the consumers. It has also been argued that unless water supply and sewage treatment services are privatized, there will not be enough of an incentive for water supply and sewerage monopolists to contain cost increases. This book aims to explore the possibility of privatization as a method of reforming the water supply and sewerage industries in Hong Kong. In order to improve the performance of the water industry, the government should consider inviting the private sector to run the industry. Hong Kong can learn from other countries about privatizing the water industry.

The structure of this book is as follows. Chapter 1 introduces the issues. Chapter 2 outlines the history of the water supply and sewerage industries in Hong Kong. Chapter 3 evaluates the performance of the water supply industry and Chapter 4 examines the experience of water privatization in the United Kingdom. In Chapters 5 and 6 the actual procedures required to privatize the Water Supplies Department are considered, on the basis of what has happened in other countries. Chapter 7 is devoted to a discussion of the privatization of the sewerage industry in Hong Kong. Instead of an outright sale of the industry, we recommend a "contracting out" arrangement. Based on the experience of contracting out the management of government tunnels, we believe that this approach will enhance competition and reduce political opposition to the privatization of the sewerage industry. A summary of our findings and recommendations is given in Chapter 8.

Note

1. "Sewerage" refers to the *system* that handles sewages and provides sewage services.

CHAPTER 2

Water Supply and Sewerage Industries in Hong Kong

History of Water Supply

Before the Second World War

Because of the absence of large rivers and underground water sources, water supply has long been a problem in Hong Kong. When Hong Kong Island became a British Territory in 1841, water supply was not a major problem because many hillside streams provided enough water for the then sparse population. Wells were sunk for city water supply, but they were often polluted by surface water and caused dysentery epidemics (Water Supplies Department 1993). As the population grew, a need for a comprehensive water scheme arose.

Such a need was first addressed by Sir John Bowring, the governor of Hong Kong from 1854 to 1859. It was his opinion that water, like all other utilities, should be provided by private enterprises. At the time Hong Kong had a primarily migrant population, and its limited number of inhabitants were mostly connected with the China trade. Hence, the financial prospects of such a private venture were not promising. The governor's idea was not adopted by the representatives of the business community in the Executive Council. The government then borrowed money from banks to build the first water supply system in Hong Kong.

The first reservoir in Hong Kong was the Pok Fu Lam Reservoir, which began operation at the end of 1863. In 1888 another reservoir was completed at Tai Tam. The Tai Tam Scheme provided Hong Kong's first supply of filtered water. The Scheme was later extended, and capacity increased significantly with the completion of the Tai Tam Tuk dam.

As available sources of water supply on Hong Kong Island neared depletion, the government started building reservoirs in Kowloon and in the New Territories. The Kowloon Reservoir was completed in 1910. Construction of the first reservoir in the New Territories, the Shing Mun (Jubilee) Reservoir, started in 1923 and was completed in 1936. By 1939 there were thirteen reservoirs in Hong Kong, and total storage capacity stood at 26.8 million cubic metres.

After the Second World War

Years of Water Rationing

For a short while following the Second World War, the government continued its efforts to increase the water supply of Hong Kong. The first project under consideration was the development of the Tai Lam Chung Valley. Supplying water in Hong Kong was undertaken by the government's Public Works Department. As the Tai Lam Chung Scheme proved to be far beyond the government's financing abilities, in 1951 the Director of Public Works proposed an interim version of the original scheme. This version was to reduce construction costs by more than 50%. The work then began and was completed in 1957.

Before the 1960s Hong Kong's water supply depended on rainfall. Shortfall in rainfall often resulted in the imposition of supply restrictions. The restrictions were for the purpose of saving sufficient water during the dry winter until the next summer rains. The water shortage problem was exacerbated by the unanticipated growth of Hong Kong's population. When Hong Kong planned its first reservoir project in 1860, the population stood at about 90,000. By the time the first reservoir at Pok Fu Lam was completed in 1863, the population had grown to 125,000, and the reservoir was too small to meet demand. Two years after the Second World War, the population in Hong Kong suddenly increased from 0.6 million to 1.8 million, and again, from 2.6 million in 1955 to 3.6 million in 1963 because of turmoil in mainland China. This unanticipated population growth put great pressure on the government to provide sufficient water.

During the post-war period there were often interruptions in the water distribution system because plants, equipment and pipelines had been seriously damaged in the war. As it took years to repair the distribution system, the government had to impose restrictions on the use of water. The government undertook the following measures as its priorities to improve the situation:

1. Rehabilitating plants and equipment as quickly as possible;
2. Developing new sources of water supply by constructing more reservoirs;
3. Improving the distribution system in urban areas; and
4. Replacing the existing stream pumps with new diesel and electric pumping sets.

A serious drought in 1954 forced the Public Works Department to restrict water use to three hours a day. When the situation became worse in 1955, the supply was reduced to two and a half hours a day. To solve the water shortage problem in some areas, the Public Works Department decided to pump water into the Shing Mun (Jubilee) Reservoir from two streams at Tai Po and also to pump water into the Kowloon Reservoir from the Shing Mun River at Sha Tin. In 1963 water use was further restricted to four hours every four days, the most serious restriction ever imposed on water use in Hong Kong (see Table 2.1).

Apart from constructing new reservoirs, the government employed drillers to sink deep wells in the hope of developing new sources of water. By February 1959, the drillers had sunk fifteen wells of which only eight were found to be productive. The eight wells supplied only a small amount of water. These poor results

Table 2.1
Water Restrictions in Hong Kong, 1954–82

Year	From	To	Hours of Supply	Total Annual Supply
1954	31.5.54	31.12.54	3 hours per day	
1955	September	October	11 hours per day	
	November	n.a.	2.5 hours per day	
1956	1.5.56	19.5.56	3 hours every other day	
	20.5.56	17.6.56	2.5 hours per day	
	18.6.56	18.7.56	7 hours per day	
	19.7.56	31.12.56	5 hours per day	
1957	1.1.57	9.5.57	5 hours per day	
	10.5.57	21.5.57	9 hours per day	
	22.5.57	Mid October	16 hours per day	
	Mid October	December	10 hours per day	
1958	1.1.58	26.1.58	10 hours per day	
	27.1.58	1.4.58	8 hours per day	
	2.4.58	9.7.58	10 hours per day	
	10.7.58	1.8.58	5 hours per day	
	2.8.58	26.11.58	10 hours per day	
	27.11.58	31.12.58	8 hours per day	
1959	1.1.59	11.3.59	8 hours per day	
	12.3.59	3.4.59	3 hours per day	
	4.4.59	23.4.59	8 hours per day	
	24.4.59	15.6.59	10 hours per day	
	17.6.59	1.10.59	17.5 hours per day	
	2.10.59	10.10.59	13 hours per day	
	11.10.59	29.10.59	8 hours per day	
	30.10.59	n.a.	4 hours per day	
1963	16.5.63	31.5.63	4 hours every other day	
	1.6.63	31.12.63	4 hours every fourth day	
1964	1.1.64	27.8.64	n.a.	3,763 hours
1966	n.a.	n.a.	n.a.	8,740 hours
1967	February	May	16 hours per day	6,062 hours
	June	End June	8 hours per day	
	1st half of July	2nd half of July	4 hours every other day	
	2nd half of July	22.8.67	4 hours every fourth day	
	27.8.67	5.9.67	4 hours per day	
	26.9.67	30.9.67	4 hours per day	
1974	25.9.74	8.10.74	16 hours per day	8,540 hours
	9.10.74	17.10.74	10 hours per day	
1977	1.6.77	4.7.77	16 hours per day	5,990 hours
	5.7.77	31.12.77	10 hours per day	
1978	1.1.78	18.4.78	10 hours per day	7,314 hours
1981	8.10.81	25.10.81	16 hours per day	
	26.10.81	31.12.81	10 hours per day	
1982	1.1.82	4.5.82	10 hours per day	
	5.5.82	28.5.82	16 hours per day	
				(Full annual supply is 8,760 hours)

Source: Census and Statistics Department

suggested that it was not worth attempting to develop new water sources in this way. The government then considered a novel idea: developing freshwater lakes.

On the northern side of Tolo Harbour, there was a large sea inlet called Plover Cove. The government's engineers conceived the idea of damming this inlet, pumping out the sea water and filling it with rainwater. Water was also pumped from other rivers into this artificial freshwater lake. The Plover Cove Scheme was a project the kind of which had never been attempted anywhere in the world. Work started in November 1960 and was finally completed in December 1968. As the demand for water continued to rise, construction work started again in 1970 to raise the height of the dams to increase the capacity of Plover Cove Reservoir. In May 1971 the government decided to go ahead with another even more ambitious project, the High Island Water Scheme. This scheme included the construction of two rock dams rising 64 metres above mean low water sea level at the eastern and western approaches of the narrow straits running between High Island and the eastern end of the Sai Kung Peninsula. The resulting freshwater lake has a capacity of 281 million cubic metres. Since the completion of the High Island Water Scheme in 1977, there have been no other major reservoir projects undertaken in Hong Kong. Table 2.2 shows the storage capacity of reservoirs in Hong Kong.

Water from China

By 1977, partly due to the difficulty in finding suitable sites, but also because by that time Hong Kong had been assured of an adequate water supply from China, the government stopped constructing reservoirs. Hong Kong started arranging to import water from China in 1960. In that year, an agreement was signed with the authorities in Po On County (Shenzhen) to supply Hong Kong with 22.7 million cubic metres of water a year from its Shenzhen reservoir, with the condition that rainfall on the Chinese side exceeded 1,600 millimetre. The Chinese continued to supply water to Hong Kong during the following two years. However, the agreement was not effective during the 1963 drought in Hong

Table 2.2
Storage Capacity of Reservoirs in Hong Kong, 1877–1978

Name of Reservoir	Year of Supply	Reservoir Storage (million cubic metres)
Pok Fu Lam	1877	0.231
Tai Tam	1889	1.490
Tai Tam Byewash	1904	0.080
Tai Tam Intermediate	1907	0.686
Kowloon	1910	1.578
Tai Tam Tuk	1917	6.047
Shek Lei Pui	1925	0.374
Reception	1926	0.121
Aberdeen (Upper and Lower)	1931	1.259
Kowloon Byewash	1931	0.800
Shing Mun (Jubilee)	1936	13.279
Tai Lam Chung	1957	20.490
Shek Pik	1963	24.461
Lower Shing Mun	1965	4.299
Plover Cove	1968	229.729
High Island	1978	281.124
Total Capacity		586.048

Source: Census and Statistics Department

Kong, as the water level in the Shenzhen reservoir was so low that it was very difficult for the Chinese authorities to release water for use in Hong Kong.

As rainfall in 1963 was extremely low and Hong Kong could not rely on water supply from China, the government's initial step was to reconstitute the Water Supply Emergency Committee, which had been disbanded after the 1955 drought. In 1963 the Committee reported to the Executive Council for all aspects of the emergency. The immediate remedies taken by the Committee during the 1963 drought were:

1. To reduce consumption by limiting water use to a four-hour period (three hours in low density areas) every fourth day;
2. To import water from the Pearl River in China by sea in tankers (with government transport arrangements);
3. To supply Hong Kong with additional drinking water from the deep tanks of normal merchant ships in the course of their regular trading voyages.

Of all these remedies, importing water from China was found to be the most effective way of tackling the water shortage problem in Hong Kong.

In 1963 the Hong Kong government approached the Chinese authorities regarding the possibility of permanently extracting water from the East River of the Pearl River system. Since the costs of transporting water via tanker shuttle service were high, the Chinese authorities advised that the most satisfactory and economical way of extracting water from the East River would be to provide dams and attendant pumps at various places along a tributary of the East River. The water would be raised in stages into a holding reservoir from which it would be discharged by gravity into the Shenzhen reservoir. The agreement was finally signed on 22 April 1964 and became effective on 1 March 1965. The project was completed in 1967. With a permanent water supply from China, Hong Kong has enjoyed a continuous supply of water since 1967, except in a few years (1974, 1977, 1978, 1981 and 1982) when rainfall was abnormally low.

The amount of water supplied by China to Hong Kong increased year by year. A China Water Supply Branch was established in February 1989 to undertake the planning, design and construction of the reception and distribution system for the supply of additional water from China beyond 1994/95. The most recent long-term water supply agreement between Hong Kong and China was signed in 1989. The agreement has ensured that China will continue to supply water to Hong Kong beyond the year 2000, with an annual increase of 30 million cubic metres — from 690 million

Table 2.3
Water Supply and Consumption in Hong Kong, 1947–95

Year (end of March)	Total Consumption (m cub. m.)	Import from China (m cub. m.)	Production by Desalter (m cub. m.)	Rainfall (mm) (end of Dec.)
1947	48			2,592
1948	53			2,483
1949	51			2,101
1950	56			2,076
1951	57			2,364
1952	55			2,544
1953	55			2,360
1954	61			1,367
1955	48			2,350
1956	59			1,649
1957	57			2,950
1958	87			2,034
1959	91			2,797
1960	102	56		2,237
1961	111	n.a.		2,232
1962	136	27		1,741
1963	119	43		901
1964	64	25		2,432

cubic metres in 1995 to 840 million cubic metres in 2000 — through a system with a capacity of 1.1 billion cubic metres per year. Water extracted from the East River is pumped over a series of dams built across the Sima River and is then discharged into the Shenzhen Reservoir before being fed by pipelines across the border at Muk Wu. After reaching Muk Wu, water is delivered along three routes to the western, central and eastern parts of the territory. At present, water from China provides about 70% of the total amount of water consumed in Hong Kong. Table 2.3 contains data on Hong Kong's water supply and consumption since the Second World War.

Table 2.3 (continued)

(end of December)				
1964	120	n.a.		2,432
1965	185	n.a.		2,353
1966	203	n.a.		2,398
1967	183	n.a.		1,571
1968	227	n.a.		2,288
1969	259	n.a.		1,896
1970	277	66		2,316
1971	301	72		1,904
1972	325	82		2,807
1973	358	84		3,100
1974	349	87		2,323
1975	361	93	8	3,029
1976	405	91	7	2,197
1977	387	131	36	1,680
1978	412	144	27	2,593
1979	467	148	0	2,615
1980	508	172	n.a.	1,711
1981	507	211	n.a.	1,660
1982	519	239	n.a.	3,248
1983	592	251	0	2,894
1984	627	285	0	2,017
1985	637	319	0	2,191
1986	703	360	0	2,338
1987	750	432	0	2,319
1988	808	515	0	1,685
1989	845	610	0	1,945
1990	873	590	0	2,047
1991	884	701	0	1,639
1992	889	663	dismantled	2,679
1993	915	627		2,344
1994	923	683		2,726
1995	919	690		2,754

Source: Census and Statistics Department
Note: m cub. m. = Million cubic metre, mm = millimetre
         n.a. = not available

Water Desalination

Another important decision in the history of water supply in Hong Kong was the introduction of desalting plants. In 1971 an experimental desalting plant, which could deal with 227 cubic metres of water per day, was commissioned. The plant was designed to test for suitable sites for large-scale desalters in Hong Kong. It was also used to test materials for the construction of desalting plants and to determine the treatment necessary to make such water suitable for injection into the distribution network. As a result of this experimental desalting plant project, it was later concluded that the most suitable site for constructing a larger desalter was at Lok On Pai, and in 1972 the government decided to construct a desalting plant there. The plant was commissioned in 1977 and was able to produce 182,000 cubic metres of fresh water per day.

Since Hong Kong had been assured of an adequate water supply from China, the government designated the Lok On Pai desalting plant into a "standby resource" on 16 May 1982. In 1989 the Executive Council decided to dispose of the plant by 1991. Because of its proximity to the new airport at Chek Lap Kok, the site of the decommissioned plant was handed over to the Provisional Airport Authority in 1992 and was then used as a transshipment centre for the construction of the new airport. The government has also commenced various major water projects for the purpose of providing water to the new airport at Chek Lap Kok and to other new developments on North Lantau.

The Water Supplies Department

Organizational Structure

The Water Supplies Department is responsible for the collection, storage, purification and distribution of drinkable water to consumers, and for the provision of adequate new resources and

Figure 2.1
Organizational Structure of the Water Supplies Department

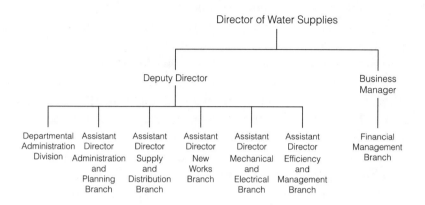

Source: The Water Supplies Department.

installations to maintain a satisfactory water supply standard. The department supplies water conforming to accepted international quality standards to the consumer and also supplies sea water for flushing purposes. The Water Authority is headed by the Director of Water Supplies, who is assisted by a deputy director and six assistant directors. These six assistant directors head five operating branches: The Administration and Planning Branch; The New Works Branch; The Supply and Distribution Branch; The Mechanical and Electrical Branch; and The Efficiency and Management Branch.

Accounting support for the department is provided by the Financial Management Branch, which is headed by a Business Manager. The Departmental Administration Division is responsible for administrative support and human resources management. In April 1996 the staff establishment of the department was 6,054, of which, 360 were professional and assistant professional officers, 2,222 technical staff, and 3,472 general and junior staff. Figure 2.1 shows the organizational structure of the Water Supplies Department.

Improving Water and Service Quality

The public's primary concern about drinking water is purity. A large quantity of the water supply is derived from protected upland gathering grounds, and pollution arising from human activities is very low. Storing the water in reservoirs for a very long period of time also permits a degree of natural purification resulting from the effects of sedimentation, aeration and sunlight. In 1939 all water in Hong Kong except in the Shouson Hill area was filtered. In the following years, more and more water treatment plants were built. By the end of 1996 there were eighteen treatment plants, with a combined capacity of 4.15 million cubic metres per day. The water quality throughout the entire supply system is continuously monitored by the Water Science Division of the Water Supplies Department and should comply with the World Health Organization's guideline values for drinking water quality.

With the expansion of the waterworks, the Water Supplies Department has introduced a computer system to assist in the operating and controlling of reservoirs as well as in the processing of hydrological data. Automated equipment has also been introduced to monitor and control the waterworks system. Since the late 1970s the department has made an effort to improve customer service. A customer relations officer was appointed to be in charge of a Complaints and Inquiries Section in the Waterworks Office. In addition, consumer enquiry centres were set up at different locations throughout Hong Kong. A customer liaison group, which includes consumer representatives, was set up in July 1993 to provide a communication channel through which customers could express their needs and their expectations of service standards. In March 1993 a performance pledge was issued and the service standards were publicized.

In the 1990s computerization and customer service continued to be the major concerns of the Water Services Department. In July 1993 a new Electronic Meter Reading System was introduced to improve the efficiency of the meter reading and billing processes. The department continues to expand its distribution network to remote areas in Hong Kong. In the outlying areas, villagers are provided with individually metered household supplies to replace public standpipes. At present, 99.5% of Hong Kong's population receives a metered water supply. The number of consumer accounts continues to rise, and the customer base had reached 2.1 million accounts by the end of 1995.

Drainage and the Environment

The Sewerage System in the Early Period

In 1881 the British government sent a consultant, Osbert Chadwick, to enquire and report on the sanitary condition of the territory. His journey marked the earliest effort to protect the environment in Hong Kong. In the early 1950s nearly all buildings were provided with their own water-borne sewerage systems. To avoid sewage being discharged into the harbour without proper treatment, a major scheme for the provision of intercepting sewers was approved in 1955. The intercepting sewers would bring sewage to selected sites where it could be chemically treated and then discharged via submarine outfalls. The intention was to make sure that sewage received at least preliminary treatment in screening plants before being discharged into the main tidal currents. The first project covered the Yau Ma Tei area of the Kowloon Peninsula and was completed in 1957.

Surface water was led to the sea through open-trained channels, known locally as nullahs. Passing down the centre of roads with bridges at road intersections, these nullahs often caused traffic congestion in some areas. From the early 1950s to the middle of the 1960s, many of these open nullahs on roads were decked to improve traffic conditions. Extensive culvert construction started in 1958 at resettlement areas to divert stream courses and allow clearance of the sites.

After serious floods in Yuen Long in 1960, the government conducted a study to investigate drainage problems in low-lying towns. A five-year project was approved to construct new culverts and to realign existing stream courses. New sewerage systems were also constructed in the New Territories. In urban areas, the Public Works Department found that the capacity of many sewers was no longer sufficient to cope with the rapid rate of redevelopment. A programme was launched to replace the old mains infrastructure with larger mains. In the 1970s the government built sewage treatment works in Sha Tin to meet the demands created by rapid development of new towns in the New Territories.

Water Pollution

Over the past two decades Hong Kong people have become more aware of the problem of pollution. The government has started to monitor the water quality in Victoria Harbour and Tolo Harbour, obtained long-term data for the design of new facilities for sewage treatment and disposal, and established pollution levels and trends. Starting in 1974 an annual data report summarizing all monitoring results has been prepared.

In the 1970s the overall problem of pollution was tackled by the Environment Branch. In 1974 the former Advisory Committee on Air Pollution and the Advisory Committee on Environmental Pollution on Land and Water were reconstituted as the Advisory Committee on Environmental Pollution (EPCOM). The committee was divided into three subcommittees that dealt separately with water and land pollution, air pollution and noise pollution. In 1977 a new Environmental Protection Unit was set up in order to develop an environmental protection policy, establish priorities and guidelines for pollution control, co-ordinate pollution control efforts, provide technical expertise on pollution control, and ensure that the provisions of environmental protection legislation were properly implemented.

The Water Pollution Control Ordinance, intended to ensure proper control of the pollution level in local waters, was enacted in 1980. The ordinance adopts an environment quality management approach that emphasizes flexibility. The water of Hong Kong has been subdivided into zones. Each zone is specified with its own water quality objective in accordance with the beneficial uses of the water in areas within the zones, such as fishing, swimming, commercial fisheries, irrigation or other activities. This is called the water control objective (WCO). The pollution control authority within the Public Works Department is responsible for implementing controls over individual discharges. In February 1982 Tolo Harbour and the Tolo Channel were declared the first water control zones (WCZ).

In 1981 the Environmental Protection Unit was transformed into the Environmental Protection Agency (EPA). The purpose of the agency was to provide a central source of expertise and scientific data on all aspects of pollution control and to assume a central co-ordinating role in the formulation and execution of government policies. A few years later, in April 1986, the Environmental Protection Department (EPD) was established. The EPD was created from the former EPA, and it carries out the work of environmental planning, the implementation of most of the pollution control legislation, and the development and review of waste treatment and disposal programmes. The department also carries out environmental monitoring and investigations, which are needed to support policy development and review. Since April 1987 all new discharges of most categories of industrial effluent and certain domestic waste water must be licensed, and the EPD is responsible for issuing licences and ensuring compliance with the license conditions. Table 2.4 shows the number of effluent discharges in water control zones (WCZ) as of the end of 1994.

The major cause of water pollution is the illegal discharge of sewage. Chemical materials used in industries, if discharged without proper treatment, can easily pollute rivers and harbours. Foul sewage outfalls illegally installed and untreated could easily lead to the contamination of coastal water. The EPD has put a great deal of effort into identifying the sources of industrial effluents discharged illegally into surface water drains entering the Tuen Mun and Fo Tan nullahs. Factory owners are required to install and use appropriate drainage facilities.

Table 2.4
Number of Effluent Discharges in Water Control Zones
(as at the end of December 1994)

Discharge	Total
Trade Effluents	
(a) Industrial	4,676
(b) Commercial	8,418
Subtotal	13,094
Domestic Discharges	
(a) Private sewage treatment plant	242
(b) Village houses and septic tanks	81,215
(c) Other premises	13,722
Subtotal	95,179
Institutional and Government Facilities	
(a) Government sewage treatment works	73
(b) Others	2,039
Subtotal	2,112
Total	110,385

Source: *Environment Hong Kong* (1995), p. 206. Reference No. 18

The Drainage Services Department

Whereas the "input" end of water works is taken care by the Water Supplies Department, the "output" end is by the Drainage Services Department. Generally speaking, the two major services provided by the Drainage Services Department are sewage disposal and flood control.

Organizational Structure

On 1 September 1989 the Drainage Services Department was set up as a multi-disciplinary department charged with the task of coming up with an efficient approach to resolving the sewage and flooding problems in Hong Kong. The Drainage Services Department is responsible for the design, construction, operation and maintenance of the sewerage, sewage treatment and disposal, and storm water drainage systems in the territory. Before the Drainage Services Department was set up, the Electrical and Mechanical Services Department was in charge of the sewage treatment works.

The Drainage Services Department is headed by a director, who is assisted by a deputy director and three assistant directors who head three engineering branches: the Projects and Development Branch, the Operations and Maintenance Branch and the Electrical and Mechanical Branch. The department is also assisted by sections rendering technical, accounting and administrative support at headquarters of the department. The division of labour is as follows:

1. The Projects and Development Branch carries out the civil engineering work associated with new storm water drainage schemes and sewerage and sewage treatment and disposal schemes.
2. The Operations and Maintenance Branch is responsible for the management of the sewerage and storm water drainage systems in the territory.
3. The Electrical and Mechanical Branch is responsible for designing and installing electrical and mechanical plants and equipment for sewage treatment plants, sewage pumping stations and flood water pumping stations.

There is also a branch called the Sewage Services Trading Fund Branch. This branch is headed by a business manager and assisted by an assistant director and other support staff for engineering and accounting services. In March 1996 the Drainage Services Department had some 1,840 employees: 218 were professional, 581 technical, 247 general supporting staff and 794 direct labour. Figure 2.2 shows the organizational structure of the Drainage Services Department.

Figure 2.2
Organizational Structure of the Drainage Services Department

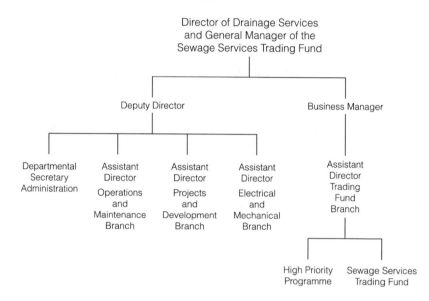

Source: The Drainage Services Department.

Sewage Disposal

Major Schemes

Sewage services are mainly provided under two major categories: construction of sewerage, sewage treatment and disposal facilities; and operation and maintenance of such facilities. At present, many projects aimed at providing sewerage and related facilities are being implemented in various stages of planning, design and construction. From a programme implementation point of view, these projects can be divided into several categories:

1. Existing Schemes
   The existing schemes are sewerage or sewage-treatment projects that had been parts of the public works pro-gramme before the current strategy to combat water pollution took shape. The North West Kowloon

Treatment and Disposal Scheme is the largest of such projects. It was put into operation in August 1992. At present, the largest project in progress is the Tolo Harbour Effluent Export Scheme, which is designed to export the sewage effluent from the Sha Tin and Tai Po sewage treatment works into Victoria Harbour and to help prevent the recurrence of red tides in Tolo Harbour.

2. The Sewage Masterplan Scheme
   This scheme includes a number of territory-wide sewerage rehabilitation and improvement projects. The territory of Hong Kong is divided into sixteen regional sewage catchment areas, each with its own treatment plant (see Figure 2.3). The scheme aims at ensuring the proper collection of sewage in foul sewers.

3. The Strategic Sewage Disposal Scheme (SSDS)
   This scheme aims at reducing the level of pollution in Victoria Harbour through a combination of land-based treatment and the natural purification process of the ocean. It is a massive project that involves collecting all the sewage from Hong Kong Island, Kowloon, Tsuen Wan, Kwai Chung and Tseung Kwan O into a deep tunnel intercepting sewer system for treatment before discharging it into carefully selected waters. There are altogether four stages under the SSDS (see Figure 2.4).

At present, much sewage in Hong Kong goes down the wrong pipes. Many people deliberately or inadvertently dispose of sewage into storm drains which are only for the disposal of rainwater. Hence, such sewage does have been properly treated before disposal. To solve this problem, the Drainage Services Department construct separate pipes for the disposal of sewage and rainwater. The department has also designated SSDS Stage 1 and six related sewage masterplan schemes as High Priority Programme (HPP) projects that are scheduled for completion in the middle of 1997. Under the HPP, sewage from six catchment areas will be collected

Figure 2.3
Catchment Areas of the Sewage Services

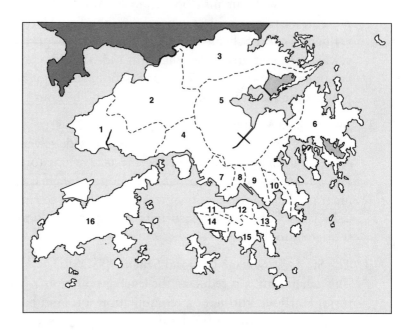

Source: The Drainage Services Department.

and screened at local screening plants and then conveyed to the Stonecutters Island Sewage Treatment Works through a network of deep sewage tunnels for chemically enhanced primary treatment.

Upon completion of Stage 1 of SSDS, 50% of the water discharged into Victoria Harbour will have been treated. The four stages of the SSDS are designed to tackle 75% of the sewage in Hong Kong. The remaining 25% of waste water will be treated by six large-scale treatment plants before discharge. The ultimate target of all the schemes mentioned above is to treat all sewage in Hong Kong before it is discharged into sea waters.

Figure 2.4
The Hong Kong Strategic Sewage Disposal Scheme (SSDS)

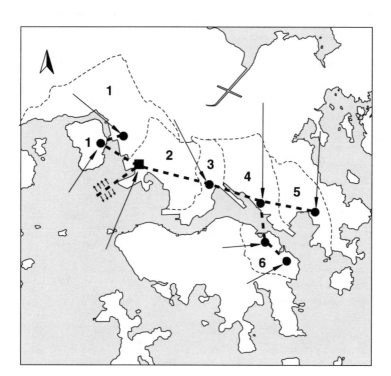

Source: The Drainage Services Department.

The Sewage Services Trading Fund

Drainage has always been an important aspect of the successful infrastructural development of Hong Kong's urban areas and new towns. In order to free Hong Kong from serious problems of water pollution and flooding, a programme of works to upgrade waterworks and drainage systems has been implemented. The total capital cost is estimated to be $22 billion. While the government is bearing the whole burden of capital cost through general revenue, a charging scheme whereby "polluter pays" has been adopted for the financing of operation and maintenance costs.

Chapter 2

Table 2.5

Sewage Charges in Hong Kong by Type of Effluent

Trade, business or manufacture	Rate for premises located in a water control zone HK$ per cub. m.	Rate for premises located outside a water control zone HK$ per cub. m.
1 Yarn sizing	3.78	10.67
2 Washing new garments, excluding laundries	0.82	0.82
3 Bleaching and dyeing of garments	0.64	0.64
4 Bleaching and dyeing of knitted fabric	1.01	1.01
5 Bleaching and dyeing of woven fabric	1.73	1.73
6 Textile stencilling and printing	1.32	1.32
7 Knit outerwear	1.01	1.01
8 Wearing apparel other than knit outwear	1.80	1.80
9 Spinning cotton	0.34	0.34
10 Laundries	0.60	0.60
11 Soap and cleaning preparations, perfumes, cosmetics	3.78	16.05
12 Medicines	3.78	4.98
13 Paints, varnishes and lacquers	1.16	1.16
14 Basic industrial chemicals	3.78	4.02
15 Tanneries and leather finishing	2.56	2.56
16 Pulp, paper and paperboard	4.09	4.09
17 Soft drinks and carbonated waters industries	1.49	1.49
18 Breweries and manufacture of malt liquor	3.29	3.29
19 Distilling, rectifying and blending spirits	0.11	0.11
20 Cocoa, chocolate and sugar confectionery	3.78	4.26
21 Vermicelli, noodles, and similar farinaceous products	3.29	5.16
22 Bakery products	3.29	5.16
23 Grain mill products	5.98	9.54
24 Vegetable oil, peanut oil, peppermint oil and aniseed oil	3.78	19.55
25 Canning, preserving and processing of fish and crustaceans	1.73	1.73
26 Canning and preserving of fruit and vegetables	3.63	3.63
27 Dairy products	3.78	9.15
28 Slaughtering, preparing and preserving of meat	3.78	9.01
29 Soy and other sauces	3.78	8.38
30 Restaurants	3.78	9.12

Source: Drainage Services Department

The concept of the sewage charging scheme is that the higher the quantity and the heavier the pollutant level of the waste water one discharges, the more one has to pay. This sewage charging scheme comprises two components: the general sewage charge and the trade effluent surcharge. The general sewage surcharge is calculated at HK$1.2 per cubic metre of water consumed. All water consumers whose premises are connected to the public sewerage system are required to pay the general sewage charges. Thirty types of trades, businesses and manufactures are required to pay the trade effluent surcharge (see Table 2.5) in addition to the general sewage charge if they generate waste water at a pollution level higher than that of domestic households. The sewage charging scheme covers not only all domestic households and industrial and commercial sectors, but also all government departments, public bodies and government premises. The general sewage charge is issued together with the water charge under the Water and Sewage Charges Bill, while the trade effluent surcharge is billed separately by the Drainage Services Department. This charging scheme covers the operational cost of government services. The cost is to be covered by revenue collected from the particular members of the public who use those services rather than by the general revenue. Government departments are also required to pay sewage charges.

A Sewage Services Trading Fund was established by a resolution of the Legislative Council in 1994. The Director of Drainage Services is the general manager of the trading fund. The trading fund is headed by a business manager and was formally established when sewage charges were introduced in April 1995. The key activities of this branch are to manage the HPP projects, to collect the sewage charges and to deliver the accounting services for the whole department.

The trading fund was introduced in the Drainage Services Department for the provision of sewage services. It covers part of the cost of operating and maintaining the public sewerage systems. The main reason for the establishment of the fund is to facilitate the "polluter pays" charging scheme. As all the sewage charges collected are paid into the trading fund, the fund enables the public to see how much it costs to collect, treat and dispose of sewage and to understand how the level of charges is determined. In addition, the trading fund arrangement provides an alternative means to getting on with work to upgrade the sewerage systems as fast as possible. The general objective of the fund is to increase the transparency of operational performance within the department, and to allow more flexibility in matching its service levels against consumer demand. The government has proposed that the trading fund should break even in five years, that is, by year 2000.

Flood Control

Flooding in Hong Kong is usually caused by the overflow of natural streams at times of heavy and prolonged rainfall. Floods occur mainly in the northern and northwestern districts of the New Territories. Unauthorized alterations in land use and redevelop- ment have also been attributed to serious flooding in some areas of Hong Kong. A plan has been drawn up by the Drainage Services Department to handle the problem. The plan consists of three components: Long-Term Structural Measures, Short-Term Improvement and Management Measures, and Legislative and Institutional Measures.

Long-Term Structural Measures

Long term measures involve the construction of major drainage works. The flood mitigation projects include two categories. The first category is made up of the main river training projects, the purpose of which is to provide a comprehensive river channel system in the New Territories to carry flood water to the sea. The second category of projects consists of village flood protection schemes for low-lying flood-prone villages, including earth bunds around the villages to protect the villagers from flood water flowing over the riverbanks, and construction of a flood pumping station to pump away the surface runoff collected within the villages.

Short-Term Improvement and Management Measures

Short Term Improvement and Management Measures are aimed at removing local drainage bottlenecks in order to prevent the flooding problem from worsening. The measures include local drainage improvement works, maintenance activities to identify and remove drain blockages, and surveillance activities to ensure that the integrity of the drainage system is preserved.

Legislative and Institutional Measures

The Drainage Services Department is carrying out Drainage Master Planning studies to review and improve the condition and performance of the existing storm water drainage system. Legislative measures have also been sought to empower the department to gain access through private land to remove obstructions in the main watercourse.

CHAPTER 3

Performance of the Water Supply Industry

Private Utilities in Hong Kong

Hong Kong's water supply has been in public hands for more than 140 years. It might be useful to draw a comparison between the performance of the water utility and that of utilities run by private business. The first private utility company established in Hong Kong was the Hong Kong and China Gas Company Limited (HKCG) which was established in 1862. The company initially contracted with the Hong Kong government to supply towngas for street lighting. Up to the present (1997), the company has not been subject to any government control on prices and returns.

In the electricity industry, the Hongkong Electric Company Limited (HEC) and China Light and Power Company Limited (CLP) are two regional monopolists that have had long histories in Hong Kong. HEC was the first electricity company in Hong Kong. It was incorporated in 1889 and began supplying electricity to Hong Kong Island in December 1890. CLP was incorporated in Hong Kong in 1901 for the purpose of supplying electricity to Canton (in China) and Kowloon (and later to the New Territories and outlying Islands). The company was wound up and incorporated again in 1918. These two electric utilities are controlled by the Scheme of Control, which was proposed by the industry in 1964. The Scheme is essentially a long-term contract (15 years) between a private firm and the government (Lam 1996).

Under the Scheme of Control, a regulated utility is subject to both (nominal) rate-of-return control and price control.

As to the telephone industry, Hong Kong Telephone (HKTC) had been a franchised monopoly on the local telephone market in Hong Kong for 70 years prior to July 1995. The company was incorporated in 1925 and was granted a licence to provide the territory with a public telephone network. Between 1983 and 1984 Cable & Wireless Hong Kong (CWHK) acquired a majority of the shares in HKTC and then merged with it to form Hongkong Telecommunications Limited (HKT). Hong Kong Telephone was under the Scheme of Control between 1976 and 1991. In 1991 the company proposed that the profit control be replaced by a price control. The government eventually accepted HKTC's proposal. Since August 1993 the company has been under price-cap regulation. The price cap for residential lines is 3% below the prevailing inflation rate measured by the consumer price index (CPI) in Hong Kong. Since July 1995 the local telephone market in Hong Kong has been liberalized, and HKTC has been competing with three other telephone companies.

Hence, electricity, gas, and telecommunications services in Hong Kong are provided by private utilities, each existing in a different regulatory environment. Should the water industry be privatized too? To answer this question, this chapter first reviews the performance of the water supply industry and then compares it with the performance of the private utilities.

Revenue and Cost Structures of the Water Supplies Department

Table 3.1* shows the revenue and cost structures of the Water Supplies Department (WSD) in the past five financial years. The revenue of the WSD is mainly derived from water charges and government contributions. About half of the total revenue in 1995–96 was derived from chargeable supplies to government and

---

* All tables of Chapter 3 are given in the Appendix starting on page 109.

non-government establishments. Contributions from government rates and from the government for free allowance to domestic consumers (for the first twelve cubic metres) provide most of the remainder of the revenue. Fees, licences, reimbursable works and interest from deposits provide less than 2% of the total revenue. In 1980–81 the share of government contribution (through rate and free allowance) was only 34% of the total revenue of the WSD (see Table 3.2). Its share has been increasing steadily and now makes up about 50% of the total revenue. In other words, water charges are being based less and less on the users-pay principle. In 1995–96, the government spent about 38% of its revenue from rates on subsidizing water.

Concerning the cost structure of the WSD, the three main cost-incurring areas are: staff costs, operating and administration expenses, and bulk purchase of water from China. Their shares in the total expenditure of the WSD in 1995–96 were 28%, 26%, and 34%, respectively (see Table 3.1). In the last five years, the staff costs of the WSD have increased at a rate faster than the overall expenditure of the department.

Performance of Private and Public Utilities

On Prices

Table 3.3 shows the water prices in Hong Kong. The average price of water, which includes contributions from the government, increased from HK$2.05 to HK$8.69 per cubic metre between 1980–81 and 1995–96. If we only consider the water charges (from government and non-government establishments), then the subsidized price of water increased from HK$1.33 to HK$4.3 per cubic metre during the same period. The rate of increase in the average price is higher than the rate of increase in the subsidized price. This suggests that the government has increased its water supply subsidy.

In Figure 3.1, we compare the average price of water with the average prices of other utility services in Hong Kong. Since data

Figure 3.1

Average Prices of Utility Services in Hong Kong, 1981–95

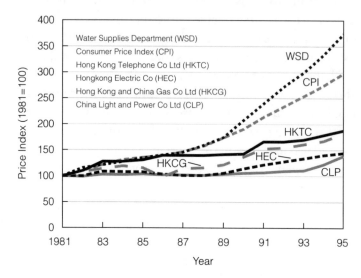

Source: Annual reports of utility companies; *Hong Kong Energy Statistics*.

from before 1980–81 are not available, we use 1981 as the base year (index = 100) on which to make our comparison. Between 1981 and 1995 the average price of water increased by 275%, against an increase in the consumer price index (CPI) in Hong Kong of 198%. In other words, the real water price has increased over time. Before 1989 the increase in water price was in general in line with inflation. After that, the price of water has been increasing at a much faster rate than inflation. Even if we consider the water price after subsidization, the price increase (209%) was still higher than the general price increase. The average prices of electricity, towngas, and local residential telephone service, on the other hand, all increased much less than the inflation rate. Their prices decreased in real terms during the same period.

It should be noted that in 1979 and 1980, the two electricity companies increased their prices rapidly to finance the construction of coal-fired generators. Even if we use 1979 as the base year, electricity prices still increased at about 2% a year less than inflation. For residential telephone lines, the price increase has been capped at 3% below the inflation rate since August 1993. The price cap of CPI – 3%, in turn, is based on the performance of Hongkong Telecom from 1980 to 1991. Water charges, however, increased above the inflation rate. In addition, the share of government subsidy in the total revenue of the WSD increased in the period under study.

As was mentioned earlier, staff costs and purchase costs of water from China make up about two thirds of the total costs. Increases in these costs would affect the price of water. Table 3.4 shows the number of employees and the labour productivity of the WSD. The number of employees in the department has been increasing very quickly as compared to private utilities in Hong Kong (see Table 3.5). Since the early 1980s, with the introduction of coal-fired generators and improvement in technology, the numbers of employers working in the electricity and telecommunications industries has been decreasing. Despite a rapid expansion in the demand for towngas since the early 1980s, the number of employees working in the towngas company has increased slowly. As a result, these private utilities have all achieved significant productivity growth.

The number of WSD staff members, on the contrary, continues to expand, even though Hong Kong has been relying more on water supply from China. The sales volume of water, likewise, has actually decreased. Using 1983 as the base year (index=100), we compare the labour productivity growth of these private and public utilities in terms of sales (or exchange lines) per worker (see Figure 3.2) and customers (or exchange lines) per worker (see Figure 3.3). The performance of the WSD is far behind that of all private utilities in productivity growth. It might be argued that the higher labour productivity achieved by other private utilities is the result of greater capital investment and technology improvement. As is clear from in Table 3.6, the government has increased its capital expenditure in the water industry, and its capital commitment will increase substantially in the future. In general, the growth of the capital expenditure in the water industry does not compare unfavourably with other private utilities in Hong Kong.

Figure 3.2

Labour Productivity (in Sales per Worker) of Public Utilities, 1983–95

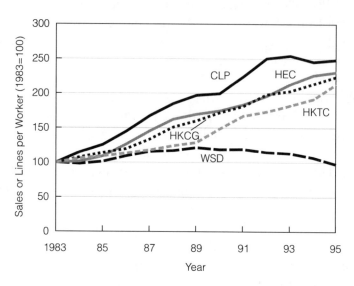

Source: Tables 3.4, 3.5 (in Appendix). For legends, see Figure 3.1.

Figure 3.3

Labour Productivity (in Customers per Worker) of Public Utilities

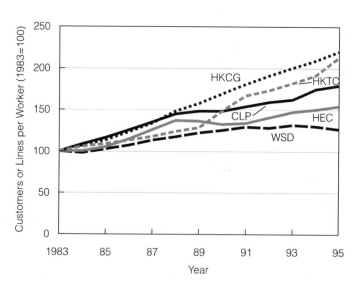

Source: Tables 3.4, 3.5 (in Appendix). For legends, see Figure 3.1.

Apart from the increase in the number of staff members, the higher price increase of water can be explained by the increase in wage rates within the WSD. From Tables 3.7 and 3.8, we can see that the nominal index of payroll per worker in the water industry has been increasing faster than that for the electricity and gas industries. Increases in the purchase price of water have also raised the cost of water (see Tables 3.9). Furthermore, as the actual sales volume of water has decreased since 1989–90 (see Table 3.4), it puts greater pressure on water charges to increase so as to generate sufficient revenue to recover the expenditure of the department.

On Returns

Although water prices have been increasing faster than prices charged by other private utilities, the earning performance of the department has been deteriorating. The actual rate of return on average net fixed assets (ANFA) of the WSD has been less than its target rate (measured by the risk-free rate of return) since 1984–85 (see Table 3.10 and Figure 3.4). In 1995–96 the department's actual return on average net fixed assets was only 3.9%. The equity returns of the WSD should be at similar levels, since the department has not issued any debt and has only borrowed a very small amount from the government to pay for water purchase.

The returns of the two electric utilities are regulated by the government. The permitted rate of return on equity-financed assets is set at 15%, while that on debt-financed asset is set at 13.5%. By debt financing, they are able to earn an equity return in excess of 20% in their electricity supply businesses. The towngas gas company has also enjoyed increasing asset and equity returns in the 1980s and 1990s. The decreases in its rates of return on asset and equity in 1993 are simply the result of the company's asset revaluation.

Since HKTC merged with CWHK and formed HKT, local telephone charges have been subsidized by revenue from international calls. Despite a subsidy provided to local telephone services, HKT has managed to earn attractive returns on total assets

Figure 3.4

Target vs Actual Returns of the Water Supplies Department, 1979–95

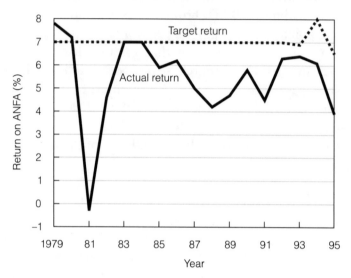

Source: Table 3.10A (in Appendix).

Figure 3.5

Water Supply and Sales in Hong Kong, 1979–95

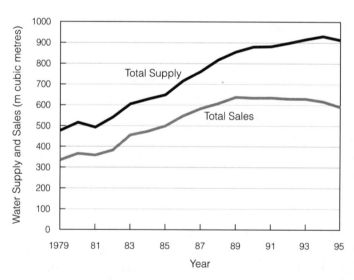

Source: Table 3.11 (in Appendix).

and equity. In 1995–95 HKT's asset return and equity return were 30% and 45%, respectively.

Other Aspects

The total supply of water in Hong Kong has been increasing steadily (see Table 3.11). Supplies are from water purchase and from reservoirs. The actual sales volume (i.e., consumption metered or accounted for) at government and non-government established can be much lower. The difference between supply and sales is due to transmission loss, free community uses (e.g., fire services) and other unpaid and unauthorized use of water. As shown in Table 3.11 and Figure 3.5, the difference between total supply and sales has been rising rapidly in recent years. In 1985–86, 77% of the water supply was accounted for; but in 1995–96 the figure had decreased to 65%.

According to the WSD, the major reason for the larger gap between supply and sales is the increase in water leakage. At present, water leakage amounts to one fifth of the total supply (*Express News,* 15 January 1997). In order to improve the situation, the department has to increase investment to renovate the existing waterpipe systems. Another source of water wastage is leakage due to public works. In 1995–96 there were more than 1,300 reported incidents of waterpipe bursts (see Table 3.12). Many of these incidents were caused by workers inadvertently breaking water pipes in roadworks. The water supply to some areas has also been disrupted for weeks because of flooding in pumping stations after rainstorms. The slow response of the WSD to all these problems has caused water wastage and has reduced the reliability of the water supply.

In the early 1970s the government decided to install a water desalter in Hong Kong. It spent billions of dollars installing the desalter, but the plant operated for only a few years. After the water supply from China was secured in 1982, the plant was left idle. The operating costs to maintain the desalter were found to be enormous (see Table 3.13). It was not until 1991 that the plant was dismantled. With hindsight, one may query the decision to invest in and to maintain the desalter, which did not really do much to help solve Hong Kong's water supply problem.

In addition, while the government recently blamed CLP's over-supply on the company's poor estimation of electricity demand, the WSD has in fact committed similar mistakes in overestimating the future demand for water in Hong Kong. Since the government has already signed a long-term water purchase contract with the Chinese authorities, Hong Kong has to increase the amount of water imported until year 2000 (Table 3.14), even though there has been a continuous decrease in actual consumption (sales) in Hong Kong.

Concluding Remarks

In terms of price, returns, and productivity, the water supply industry in Hong Kong compares unfavourably with private utility companies. The financial performance of the WSD has been deteriorating in recent years. The department has not met its target return since 1984–85. With decreasing water sales and continuously rising staff and water purchase costs, water prices have risen above inflation. The government has had to increase the subsidy provided for water supply. Despite the price increases, there are signs to indicate that labour productivity, water quality and supply reliability are deteriorating. The problems of water leakage and over-supply have been becoming more serious. Increases in capital investment are required to improve the water supply systems.

In order to improve the performance of the water industry, the government should consider inviting members of the private sector to run the industry. Hong Kong can learn from other countries about the privatization of the water industry. In the next chapter, we will discuss the experience of privatization in England and Wales.

CHAPTER 4

Water Privatization Experiences

The U.K. Water Industry

Before Privatization

The pre-privatization structure of the water industry in England and Wales was established in 1973. Ten Regional Water Authorities (RWAs) were created by the 1973 Water Act. The Act unified a highly fragmented water industry in which water supply, sewerage and environmental services had been distributed among multifarious organizations at a local level (Hunt and Lynk 1995). The principle behind this reorganization was called Integrated River Basin Management (IRBM), whereby individual RWAs controlled and planned all water usage in each river catchment area. In addition to their commercial functions of water supply and sewerage, these ten RWAs were charged with the regulatory functions of monitoring the quality of water and protecting the environment. The reorganization was aimed at achieving the economies of scale and scope which were believed to be available under a vertical structure of larger integrated authorities. However, the combination of regulatory and commercial functions within the same organizations meant that these RWAs were both "poachers" (polluters of rivers with effluent discharges) and "gamekeepers" (regulators of water quality). It was therefore possible that their separate objectives would conflict with each other (Armstrong et al. 1994).

In the 1980s the ten RWAs were subject to three external disciplines: investment, borrowing, and operating costs. In the case of investment decisions, RWAs were required to earn a target real rate of return on new investments and a lower but increasing rate of return on the current value of existing assets. Each RWA's borrowing was also controlled by an external financing limit; any investment above this limit had to come from retained earnings. In addition, there was a "performance aim" that specified a target figure for operating costs. Throughout the 1980s the controls imposed on investment and external borrowing were progressively tightened. As a consequence, water charges had to be increased to provide enough internal funds for investment. In spite of rising water charges for financing new investment, there was a general belief that there was insufficient investment during the 1980s.

Apart from the ten RWAs, there were other privately owned statutory water companies that did not have sewerage functions. These water companies acted as agents of the authorities and supplied water to about one quarter of the population of England and Wales. They were under a regulatory regime different from that governing the ten RWAs. They were subject to a form of dividend control. The government also controlled their dividends, transfers to reserves and accumulated reserves. Subject to those limits, the companies were free to increase prices without additional approval.

The internal efficiency of the RWAs had been judged inadequate by the U.K. government, and the 1983 Water Act reorganized their management boards with the aim of introducing a more commercial and cost-conscious approach to their operations. Three years later a White Paper on the privatization of the RWAs was published. The White Paper recommended that the ten RWAs should be privatized in their existing form, with both commercial and regulatory functions and with vertical integration of water supply and sewerage services. The White Paper argued that "the catchment-based structure of the water industry has worked well in practice. It has been recognized throughout the world as being a good and cost-effective model for other countries to follow . . . . The water authorities' ability to operate on the basis of integrated river-basin management and to plan and develop water resources regionally has enabled them to improve their services and to keep pace with rising demand." (The U.K. Department of the Environment 1986)

The Department of the Environment invited Professor Stephen Littlechild of Birmingham to examine the form of regulation imposed on privatized water authorities. Littlechild recommended a price-cap rather than a rate-of-return regulation, as a price-cap regulation would be simpler and could preserve the efficiency incentive of the regulated firm. He had made the same recommendation when British Telecommunications (BT) was privatized in 1983 (Littlechild 1983). Littlechild also argued that the existence of ten RWAs would widen the scope of competition by enabling the market to compare the operation and performance of different authorities. The take-over threat mechanism would force companies to improve performance.

Although the 1986 White Paper recommended that the British government maintain the existing vertical structure of the RWAs, the Department of the Environment eventually decided to abandon the principle of the IRBM by setting up the National Rivers Authority (NRA) in 1987. The NRA is responsible for the control of water pollution and the management of water resources. The government discarded the idea of a wholesale transfer of the RWAs to the private sector in the belief that a privately owned water industry would not effectively regulate its service quality. Hence a separate body was needed to impose environmental regulations and to set standards of water quality.

After Privatization

The framework of the privatization of the U.K. water industry in England and Wales was formalized by the 1989 Water Act. Licences were issued to the ten RWAs and other water companies. The ten RWAs are now commonly referred to as water and sewerage companies (WASCs), while the previous statutory water companies are known as water only companies (WOCs). At the time of privatization there were 29 WOCs; since then their number has been reduced by mergers. The 1989 Water Act provided for the establishment of the Office of Water Services (OFWAT) headed by a director general. Ian C. R. Byatt is the first director general. The duty of the director general is to ensure that water and sewerage operations are properly carried out and that companies are able to earn reasonable returns on their capital so that they can finance the proper running of their operations. Since the privatized companies are in many respects monopolists, and since there are few "pure" market pressures on them, OFWAT has to assume responsibility for protecting consumer interests (Byatt 1991).

Since privatization, charges by WASCs and WOCs have been controlled by a price cap in the form of RPI + K where RPI is the inflation rate measured by the retail price index. The K factor in the formula is an allowable price increase above inflation to be used to finance the investment plans necessary to upgrade capacity and meet quality standards. This contrasts with other privatized utilities, most of which are subject to the RPI – X price-cap formula. Five services are included in the price cap. Each company was originally given a specific K profile for ten years, and a review of the regulatory formula was to be conducted after ten years. Both the regulated companies and OFWAT, though, had the right to bring forward the review from 1999–2000 to 1994–95. In the first full year after privatization (1990–91), the average K for WASCs was 5.35% (with a range of 3% to 7%), while the corresponding average for the WOCs was 11.4% (with a range of 3% to 25%). The K profile for each company is designed to reflect the company's need for financing investment. Hence, each company is able to plan its investment programme in a reasonably stable environment.

In order to fix the K factors, the regulators have first to determine the cost of capital and the value of the existing asset base so that water companies are able to earn reasonable returns on their capital. By employing the capital asset pricing model (CAPM), the Department of the Environment estimated the real cost of capital for new investment at 7% for WASCs and 8% to 8.5% for WOCs. Concerning the value of existing assets, it was decided that they should be valued in such a way that the existing owners would not gain or lose from the change in the regulatory regime. The term "indicative value" (discussed in Chapter 6) was created for the purpose of valuing existing assets. Before privatization, the WASCs were earning rates of return on the replacement cost of assets of about 2%. Thus, instead of raising the rate of return to 7%, the low 2% rate of return was maintained for the existing assets until they were fully depreciated.

Evidence has shown that both water supply and sewerage functions have enjoyed substantial economies of scale and economies of scope. The rate of technological change in the water industry is slow, and there are no significant opportunities for product market competition. When privatizing the water industry, the British government relied on a system of "comparative" or "yardstick" regulation. By comparing the performance of the 39 water companies and using the best to set standards for the others, comparative competition was introduced into the regulatory regime. In setting the standards, the regulator considers the differences among companies in operating costs, capital costs and levels of service. There are also allowances for differences such as geographical conditions that are outside the control of efficient management (Sawkins 1995). Although there have been substantial problems in implementing such a system of yardstick competition in the U.K. water industry, Sawkins's 1995 study concludes that the system will be workable after the regulator has developed more sophisticated and reliable methods for making comparisons and establishing controls for company heterogeneity.

Figure 4.1 shows the relationships between the core regulators in the U.K. water industry. An approach that separates economic regulation from environmental regulation is adopted. There are some advantages arising from adopting a separation, rather than an integration, approach to regulation. First, we can reap the benefits arising from specialization in regulatory duties. Second, different regulatory bodies can exert a positive check on each other, thereby

Figure 4.1
Relations among the Core Regulators in the U.K. Water Industry

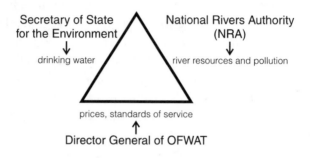

Source: Byatt (1991), p. 165.

reducing the possibility of regulatory capture. However, conflicting regulatory objectives may be found in the separation approach. For example, if the environmental regulator tightens quality standards, then there will be pressure on price increases, which may not be allowed by the economic regulator. Progressive tightening of environmental standards will result in higher and higher charges, which may be in conflict with the economic regulator's objective to protect consumers.

The Privatization Experience

The First Few Years

Since the U.K. water industry was privatized in 1989, there has been increased public dissatisfaction regarding pricing, charging structures and the conduct of the regulators. Increases in water prices have been ahead of inflation, while the rates of return earned by the privatized water companies have been high. The recession in the early 1990s has brought about a collapse in the construction industry. The subsequent reduction in construction costs in the industry has increased the returns of water companies under the RPI + K formula. To preserve the incentives for water companies to reduce costs, the economic regulator, OFWAT, has tightened the price-cap formula only slightly. Even though the privatized water companies have realized windfall gains from falling construction costs, they have to face a regulatory risk — that production costs might increase when the quality regulator imposes more stringent controls on environmental standards.

Because of the general squeeze in public expenditure before privatization, there was a need for the privatized water companies to increase investment to renew infrastructure. In addition, more stringent environmental standards imposed by the European Commission have forced the water companies to increase their levels of quality-driven investment. Between 1980 and 1993 investment in the water industry accounted for 3% of gross fixed capital formation in England and Wales (Helm and Rajah 1994). Hence, one direct consequence of privatization is that the water companies have effectively taken on a large segment of capital expenditure and freed the government from financing the investment through general taxation.

As a result of the separation of environmental regulation from economic regulation, the problem of conflicting objectives has been a persistent feature of the water industry since privatization. Discussions have been held between the NRA, OFWAT, the water companies and the Department of the Environment in the hope of providing a channel for co-ordination of environmental and economic regulations. The Director General of OFWAT has also stressed the importance of customer consultation so that customers can reveal their preferences regarding water quality and the prices they are willing to pay for it. The Director General has argued that "unless the customers feel they are getting value for their money, the whole regime could be at risk" (Booker 1995). Some analysts have argued that the problems arising from the separation will only be fully resolved by once again integrating the two regulatory activities.

In addition to debating the issue of the separation of environmental regulation from economic regulation, critics have argued that privatization has introduced an industry structure inappropriate to the natural monopoly status of the water industry.

The argument focuses on the loss of economies of scope arising from the joint supply of water, sewage, and sewerage and environmental improvement due to the transfer of the latter function to the NRA. Others argue that privatization has been attempted in the absence of sufficient evidence demonstrating the superiority of private ownership over public ownership. Lynk (1993) has measured the efficiencies of the private (statutory) water companies and the 10 public RWAs in the pre-privatization water industry and concludes that the average level of efficiency of the latter was higher in the period immediately preceding privatization. But he notes that the improvements in productivity within the public RWAs were induced by the target reductions in unit operating costs imposed by the government. Lynk also argues that the transfer of the RWAs to the private sector, coupled with the creation of the NRA, has caused a loss in efficiencies of joint production. Hunt and Lynk (1995) argue that the lost economies of scope constitute a notional "price" that consumers have paid for the creation of the NRA so as to avoid a self-regulating privatized water industry.

The Periodic Review

Taking into account the problems of conflicting regulatory objectives and the substantial changes in the operating environment of the water industry, the Director General of OFWAT decided to use his discretion to initiate a periodic review in 1994, five years after the onset of privatization. One way in which the regulatory structure was possibly to change was by integrating economic and environmental regulation with the creation of a unified regulatory body for the water industry. Another major change to take place was the introduction of a new price-cap formula: $RPI - X + Q$. The Q factor in the new formula reflects the increase in capital expenditure required to meet environmental standards, while the X factor is the amount that can be saved by greater efficiency. It is hoped that the new formula will provide a satisfactory way to resolve the trade-off between higher quality and higher charges.

As part of the 1994–95 Periodic Review, the water companies had to draw up a programme for investments covering all areas over the next 20 years. This programme was called the asset management plan and was part of an overall Strategic Business Plan submitted to OFWAT. The Director General of OFWAT considered these investment plans when setting the Q factor in the new price-cap formula. In estimating the cost of capital, OFWAT considered two models: the capital asset pricing model (CAPM) and the dividend growth model. In the former model, the cost of equity is calculated as the sum of the risk-free rate of return and the risk premium for a particular company relative to the market risk. The risk premium for a company is captured by the beta value, which measures the risk to the investors when holding the stock of the company. Another method, the dividend growth model, measures the cost of equity as the sum of dividend yield plus dividend growth.

The Director General of OFWAT has argued that the water industry is, in many respects, a relatively low-risk industry. Holding stock in a water company is similar to holding a bond, which provides a relatively stable and certain stream of income. He has proposed a relatively low cost for equity capital. Inevitably, this proposal has received strong opposition from the water companies. They have argued that subject to an uncertain regulatory regime, the risk faced by the water companies is relatively high and should be compensated for by a relatively higher rate of equity return.

The U.S. Water Industry

The water industry in the United States is very fragmented and pluralistic, as is its regulatory process (Beecher 1995). A large number of different water systems are regulated in different ways. State regulators are responsible for implementing water quality standards promulgated by the U.S. Environmental Protection Agency (EPA). Many water utilities are also subject to regulation by interstate or intrastate organizations. These regional regulatory bodies have substantial authority over the operations of water utilities.

The majority of water utility customers are served by publicly owned water utilities. Privately owned water systems account for 15% of the water industry (Williams 1992). Most of these private systems are regulated by the state public utility commissions because of their natural monopoly status. It has often been argued that the monopolistic character of water utilities undermines opportunities for competition. On the other hand, water supply technology and cost characteristics would limit many forms of competition. Water should be supplied through a vertically integrated water utility, as economies of scale and economies of scope in water supply development and treatment are substantial. The rate-of-return regulation is therefore imposed on these monopolistic water utilities. In recent years, following the global trend, there has also been an increasing interest in the privatization of water utilities in the United States.

There has long been a debate over the relative efficiency of public versus private enterprise. The basic argument is that the lack of transferability of property rights in a public enterprise eliminates the incentive element of ownership and inhibits the capitalization of efficient decisions into the present value of the firm (Bhattacharyya et al. 1994). In the Bhattacharyya study, they show that both public and private water utilities in the United States exhibit significant relative price inefficiency in addition to excessive capitalization. They also find that public water utilities are, on average, more efficient than private water utilities, even though they have a wider dispersion of performance. One of the reasons for the relative inefficiency of private water utilities is that they are relatively burdened with the cost of regulation. Rate-of-return regulation imposed on water utilities can cause allocative and dynamic inefficiency. As the rate of return is fixed by the regulator, the incentives of private firms to improve productivity is lowered. In addition, a large amount of money has to be spent in the regulatory process. Concerning the pricing of privately owned water utilities, an empirical study conducted by Williams (1992) shows that they set rate structures in favour of industrial customers at the expense of residential and commercial customers.

Privatization: Benefits and Costs

In the past two decades, there has been a global trend towards the privatization of public enterprises. This international wave of privatization is considered to have begun in the United Kingdom and spread to other industrial and developing countries (Cook and Kirkpatrick 1995). The pace of transferring public enterprises to private ownership appears to have accelerated in recent years. Privatization of the U.K. water utilities occurred in the 1980s and the associated benefits and costs are briefly reviewed below.

The Benefits of Privatization

The privatization of water utilities in many countries often took place after cutbacks in the levels of government investment and service, resulting in poorer quality of water services. In the United Kingdom new investment was required in the 1980s to improve infrastructure in the water industry and to meet new environmental standards. It was perceived that privatization can free the managers of water utilities from the financial restrictions imposed on them by governments, providing them with the flexibility to raise funds to finance expansion and improve efficiency. It was also anticipated that the government could obtain revenue from the proceeds of the sale of public enterprises.

Financial benefits aside, water utilities would gain a sense of initiative and independence after privatization. They would also gain a greater incentive to increase productivity and diversify into new businesses in order to increase their returns. Finally, accountability would also increase because, faced with direct or indirect competition, privatized water companies would recognize the importance of meeting the needs of their consumers. By and large, the experiences of the British and American water utilities

have borne out all of the above benefits except for specific problems
in minor respects.

The Costs of Privatization

A move to private ownership necessarily means that water
companies would give their shareholders priority over consumers.
Hence, new regulations may be required to prevent monopolistic
behaviour on the part of the water companies. One risk of
privatization is that instead of protecting the interests of consumers,
the regulator could be "captured" by producers to design policies
favouring the industry. Additional governance costs would
therefore have to be incurred in order to monitor the behaviour of
the privatized water monopolists and to prevent regulatory capture.

Privatization of water utilities would quite probably result in
substantial increases in the prices of water services. This is because
newly privatized firms must increase investment to improve
infrastructure and to meet specific regulatory obligations in respect
to environmental and community protection. A case in point is that
the U.K. water industry required a large amount of investment after
privatization as a result of previous cutbacks in government
investment. This led to rapid increases in the price of water services
in the early 1990s. Such hikes in prices were simply the result of
increases in quality standards (imposed by the government), rather
than the result of privatization. In Hong Kong, too, water prices
will increase if the water industry is privatized. This is because once
the industry is privatized, the existing government subsidy will be
removed and additional investment will be made to improve service
quality.

Another cost concern is that the privatization process will cause
increases in capital costs. In various latent forms, such a risk is
inherent in the process. Following privatization, the water
companies would no longer be under government protection. As
the government would bear no responsibility for financing the
provision of water services, the privatized water companies have to
generate sufficient business income to cover their operating

expenses. Moreover, when these companies diversify to other non-utility businesses, they will face higher risk in the operations of such businesses. With privatization, water companies must also face a regulatory risk arising from the uncertainty of future regulatory obligations. This regulatory risk is particularly high for the water industry because new capital is often required to renew infrastructure and to meet new environmental obligations.

On balance, the benefits of the privatization of the U.K. water industry have been found to outweigh the costs (Meredith 1992). However, there has been increasing public concern over the distribution of the benefits. In spite of the recession in the United Kingdom, water company shares outperformed the market by well over 30% in 1992 (*Investors Chronicle*, 7 May 1993). On top of worries about water quality and the environment, customers are increasingly concerned about their water bills (*Institutional Investor* 1992). The merits of privatization having been settled, current debates seem to have shifted toward the problems with the distribution of the potential gains.

CHAPTER 5

Privatization of the Water Supplies Department

Introduction

The Hong Kong public has been dissatisfied with the services provided by the Water Supplies Department. Compared with infrastructure services provided by private utilities, the transmission network developed by the Water Supplies Department seems to be unable to meet the public's expectations. As living standards improve, people will inevitably expect cleaner water to drink and receive better service. Over the past few years, the problems of poor water quality and supply disconnections in certain districts in Hong Kong have drawn severe criticisms. The Water Supplies Department's slow response to water leakages and disconnections after natural disasters (e.g., typhoons and flooding) has resulted in mass dissatisfaction. Although a declaration of performance pledge has helped the Department to improve performances, water services still compares unfavourably with those of other private utilities.

After years of study, the government is considering establishing a trading fund for the Water Supplies Department. It is expected that the revenues derived from the provision of water services will cover the operating expenses of the department, and that there will be a reasonable return on assets. The government has set the target rate of return at the risk-free rate of 6.5% to 8% (Finance Branch 1994). However, as we have seen in Chapter 3, the Department has not been able to earn this risk-free rate in the past decade. Since the

Figure 5.1
The Proposed Industry and Regulatory Structures of
the Privatized Water Industry in Hong Kong

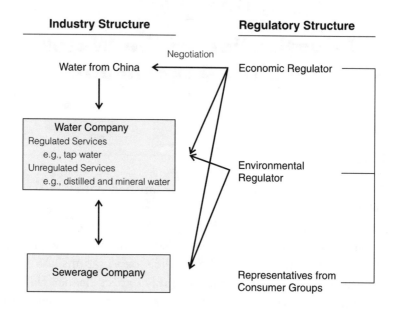

Department can always rely on government subsidy, it has little incentive to cut costs to improve its financial performance. Water Supplies Department does not enjoy price flexibility because price changes have to be approved by the Legislative Council whose members have tried to block price increases on the ground that water is a daily necessity and must therefore be subsidized by the government.

Price increases have become politicized issues. Share prices of public utilities have fallen relative to other stocks due to introduction of greater competition into the public utility industry. Utility companies must seek ways to improve profit prospects, for example by improving productivity and decreasing production costs. Historically, water services were supplied by the government, and hence the market mechanism to induce productivity improvement does not exist. In the late 1980s the government decided not to privatize the water supply industry for security reasons (*Sing Tao Jih Pao*, 8 August 1987). There are lessons to be learnt from the experience of privatization in other countries over the last decade, and the government should now reconsider the option of privatizing the water supply industry.

Industry and Regulatory Structures

In Hong Kong, two different departments, the Water Supplies Department and the Drainage Services Department, are respectively responsible for water and sewerage services. Another department, the Environmental Protection Department, monitors water quality. If the government decides to privatize the water supply industry, two classes of issues must be addressed: first, the structure of the industry and second, the regulatory structures of the newly privatized industry. Figure 5.1 shows the proposed industry structure and the regulatory structure of the privatized water industry in Hong Kong.

According to the proposed trading fund arrangements for the two departments, customers should bear the capital costs of *water* services, but the government should bear all the capital costs of *sewerage* services. Although there are obvious economies of scope from merging the two services, we recommend that the existing structure of vertical separation be maintained. In Chapter 7 we will consider some of the economic arguments for the government to bear the capital costs of providing sewerage services and flood control. In the case of water services, however, there is no special reason for exempting customers from the capital costs. It is difficult to establish any sound economic reason for government subsidization of the capital costs of water services. In many ways, water supply is no different from other utility services and so water utility should also operate on a commercial principle and that prices charged by the water utility should cover capital costs.

Slow technological development and the small economies of scale have limited the opportunities for market competition in the water industry. Yardstick competition is infeasible in Hong Kong given its small supply area. The privatized water industry is therefore likely to be a natural monopoly. Hence, government regulation would still be needed to protect the customers. The main question is: Should the U.K. approach of separating economic regulation from environmental regulation be followed by Hong Kong?

If such a separation is desired, then a mechanism should be built into the system to address the problem of the trade-off between water quality and water charges. Such a mechanism would also aim to keep the regulator and the privatized water utility better informed regarding consumer preferences and the trade-off between price and quality. The economic and environmental regulators would have to search for ways to involve customers in determining pricing and levels of service. We could expand the roles of the existing Customer Liaison Group established by the Water Supplies Department in 1993 to cover matters including water charges and quality expectations. Through the process of customer consultation, the regulator would act as an agent of the consumers to monitor the price and quality of the privatized water services. Formal communication channels between economic and environmental regulators would have to be established. In addition, the regulatory responsibilities of the two parties should be clearly defined and governance costs should be kept minimal.

Necessary Steps Before Privatization

Injecting Financial Resources and Improving Productivity

After water industry is privatized, water charges almost always increase. The increase is necessitated by the new investment. One way to avoid political opposition to privatization would be to keep water prices low during the period *preceding* privatization. Before handing the industry over to the private sector, the government should consider injecting sufficient funds into the industry so as to meet future capital requirements. Alternatively, the regulator should make clear to the public about the water company's business plans before its sale to the private sector. The water utility should submit a long-term investment plan in order to better inform the public about the expansion of the company and the implications for prices and costs.

Before privatization, the government of Hong Kong should find ways to enhance the productivity of the Water Supplies Department. One immediate consequence of such an attempt could be massive layoffs, especially at the lower-ranking level (*South China Morning Post*, 20 September 1995). The government would have to design an appropriate redeployment scheme and compensation package. If productivity has already increased, the water monopolist would have less opportunity to raise returns by further cutting manpower after privatization. This would allow for more accurate projections of future prices and profits, thus avoiding consumer dissatisfaction over excessive profits earned by the water monopolist. Furthermore, if consumers see rising investment, falling prices, and improved productivity before privatization, they would develop a more favourable attitude towards privatization of the water utility. In the future, the lessons learnt from the privatization of the water industry could be applied to the privatization of other public enterprises in Hong Kong, such as the Mass Transit Railway and the Kowloon–Canton Railway.

Changing the Water Charging System

In England and Wales, only a small proportion of domestic water consumers have meters installed, and the charging system is based on the rateable value of the property. OFWAT has consulted widely on metering, and legislation has been formulated to require the U.K. water industry to find an alternative charging system. In Hong Kong, the situation is just the opposite; nearly all water customers have meters installed. An "increasing block-pricing structure" is applied to domestic users, whereas a single unit price is adopted for non-domestic users (see Table 5.1). At present, revenue from water charges cannot cover all production costs, and the government use part of its revenue from rates to subsidize water services. Non-domestic users enjoy a larger subsidy than domestic users.

Table 5.1
Water Charges in Hong Kong (since 16 February 1995)

1. Domestic supply rate for a four month period

		per cubic metre
First tier:	First 12 cubic metres	Free
Second tier:	Next 31 cubic metres	$4.16
Third tier:	Next 19 cubic metres	$6.45
Fourth tier:	For the remainder	$9.05

2. Fresh water for flushing for a four month period

		per cubic metre
First tier:	First 30 cubic metres	Free
Second tier:	For the remainder	$4.58

3. Non-domestic supply rates

	per cubic metre
Trade purposes	$4.58
Construction purposes	$7.11
Ocean-going shipping purposes	$10.00
Non ocean-going shipping purposes	$4.58

Source: Water Supplies Department
Note: Dollar amounts are in HK$.

Table 5.2
Average Monthly Household Expenditure on Utility Services in 1995

	HK$
Water Supply	38
Sewage Service	12
Electricity (HEC)	355
(CLP)	242
Towngas	156
Telephone	63

Sources: Waters Supplies Department of Hong Kong Government, HEC, CLP, HKCG,
          and HKTC
Note: Dollar amounts are in HK$

There is a cross-subsidy from domestic to non-domestic users. As the payment of rates is not directly related to water usage, there is also a cross-subsidy from industries that are not water intensive to those that are.

The increasing block-pricing structure is designed to encourage conservation. Unfortunately, charging water below cost encourages waste because there are subsidies from rates. Charges should be based on consumption and production costs if we want to achieve a more efficient use of water. To accomplish this goal, the government should consider rebalancing the existing tariff structures: water charges for all users (particularly for non-domestic users) should be increased, and the rates paid by property owners and users should be reduced accordingly. To encourage water conservation, the government may continue to maintain the present free allowance provided to small users.

According to our estimation, water prices have to be doubled if the subsidy from government rates is to be eliminated. Table 5.2 shows the average monthly household expenditure on utility services in 1995. The expenditure on water and sewage services is the lowest among all the utility expenditures. Even if water prices are doubled, household expenditure on water will still be much lower than electricity and towngas. Given that the government continues to provide a free allowance to small users and that rates are to be reduced accordingly, we believe that this 100% price increase should be economically affordable and politically acceptable to the public.

Controlling Monopolistic Behaviour

After privatization, a form of price control may be required to prevent monopolistic behaviour and to enhance improvement in productivity. This form would include three elements. A price-cap regulation would be implemented to encourage the privatized utility to earn a reasonable return on new investment (the K factor) and to reward efficient production (the X factor). The price cap can contain an automatic adjustment mechanism to allow for cost changes that are beyond the utility's control, for instance, price hikes imposed by China which supplies about 70% of Hong Kong's water in 1996. The Hong Kong government can continue to negotiate the terms of the supply contract with the Chinese authorities. The annual adjustment in the price of this water source from China's East River can be incorporated in the price-cap formula (the Y factor). Hence, the price-cap formula can take the following form:

$$CPI + K - X + Y,$$

where

  CPI = percentage change in consumer price index;
  K = a factor that reflects the percentage change in price required to finance expansion;
  X = a productivity factor (as a percentage);
  Y = cost pass-through.

Any change in the purchase price of East River water would be automatically passed through to customers. The Y factor in the above formula is similar to the automatic fuel adjustment clause adopted by electricity companies in Hong Kong; it reflects the effect of a change in the factor price on the overall price level.

Allowing Business Diversification and Competition

In order to enhance the profit prospects and to attract private interest in the water supply business, the government may consider allowing the privatized water utility to diversify into related businesses. As living standards in Hong Kong improve, demand for value-added water products like distilled water and mineral water is increasing. The water utility can enter into this expanding market and compete with other water suppliers.

Another business opportunity lies in co-operating with China to construct reservoirs in China. Land area in Hong Kong is limited, and it is not easy to find suitable sites for new reservoirs. Although the Chinese authorities have assured Hong Kong's water supply until year 2000, there have been reports that urbanization and industrialization along the East River have caused water pollution problems and that such developments have reduced the available water supply. Apart from changing the water charging system to encourage conservation, we have to seek alternatives ways to increase water supply beyond year 2000. Constructing new reservoirs in China can bring benefits to both Hong Kong and China. While China can provide the land and labour needed in the construction of reservoirs, Hong Kong can provide the engineering support and management skills. This will effectively utilize the resources of both partners.

For many years Hong Kong has been relying on China for water, and small reservoirs in Hong Kong may not be as useful for providing water as before. These reservoirs, together with the associated catchment areas, can be redeveloped for other purposes (*Hong Kong Economic Journal*, 2–3 December 1996). In Hong Kong, the Wong Nai Chung Reservoir Park was converted from a disused reservoir under the same name. Because of its small capacity, the reservoir was converted and handed to the Urban Services Department for recreational purposes in 1978. The government may consider allowing the privatized water utility to co-operate with other private business enterprises to redevelop some reservoirs and catchment areas.

Hong Kong's reliance on the East River for its water has subjected itself to China's monopoly pricing. In the future, the privatized water utility should be encouraged to seek new sources of water in China. The government can modify the cost passthrough factor in the price-cap formula to induce the privatized company to seek cheaper sources of water. Hence, a privatized water industry may facilitate the development of a competitive market for water in mainland China.

Concluding Remarks

The benefits of privatization of the water industry are found to outweigh the costs. Based on U.K.'s privatization experience, two concerns are expected to be raised: how the realized benefits should be distributed, and how much consumers are willing to pay for higher-quality water services. Hence, the government should establish some consultative mechanisms to reveal consumer preferences. To avoid consumer dissatisfaction over excessive profits earned by the privatized water utility, the government should take necessary steps to improve the charging system and to enhance the productivity of the company before privatization. Once the utility is privatized, further improvement in efficiency to raise profits should be made.

On the investment side, the government can do either one of two things. It may inject enough investment funds into the utility. It may allow prospective investors to be clearly informed of the company's future expansion plans so that long-term development of infrastructure in the water industry will not be adversely affected. To encourage investment in the industry, the water utility's price charges should cover the costs of capital. In addition, the privatized company may be allowed to diversify and to find other ways of enhancing the company's business prospects. Once the investment plan, pricing structure and business strategies of the company have been determined, prospective buyers will have a better sense of the worth of water company shares. In the next chapter, we will consider the procedures that must be followed in the sale of a government asset.

CHAPTER 6

The Sale of a Publicly Owned Water Utility

Asset Valuation

One of the most important steps for selling an asset is to assess its value. Setting too low a price would result in an unnecessary windfall gain to the purchasers and would reduce the proceeds received by the seller; but setting too high a price would render the sale unsuccessful. The selling of a public asset is even more problematic than that of a private one, as the former seldom has a previously quoted value, and it is difficult to find a similar marketable asset for comparison of value. In the United Kingdom, it has been observed that many public assets were sold at a discount during the privatization process, and that the stock price of these enterprises rose tremendously once they were listed on the stock exchange.

Book Value, Replacement Value and Indicative Value

There are several valuation methods to set the price of an asset when it is sold to a third party. For example, the asset may be priced at its book value, at the replacement cost, or at the "indicative value" which is a concept adopted by the British government. The book value of an asset represents its historical cost minus any depreciation that has occurred. In the case of the Hong Kong Water Supplies Department, all its fixed assets are stated at their costs of acquisition except for land and capital projects (Treasury 1996). No cost is included for land that is occupied by installations or sterilized by catchment areas. For capital projects, the costs include the actual direct expenditure, and staff costs for the design, planning and supervision during the construction period. Depreciation of fixed assets is calculated on a straight line basis to write off the costs less residual value over their estimated useful lives. The annual rates of depreciation are different among various types of assets, depending on their nature. For example, more durable assets such as tunnels and dams are calculated to depreciate at 1% annually, while mechanical or electrical works, meters and plants are set to depreciate at an annual rate of 4%. The term Net Fixed Assets (NFA) used by the Treasury of the Government is equivalent to the book value of fixed assets.

The replacement cost of the asset refers to the amount that must be expended in order to acquire that asset on the market. In the case of the fixed assets of a water utility, the replacement cost equals the total costs that would be incurred if one wanted to rebuild the water transmission networks and reservoirs as well as to purchase the capital equipment at the current market price. In order to compute that replacement cost, we need to spend a significant amount of time in estimating the resources for rebuilding the many large-scale and complex projects owned by the Water Supplies Department.

It is inappropriate to adopt either the book value or the replacement cost as the valuation method for privatizing the water utility. The book value only accounts for the historical cost figures, while the replacement value of an asset is a forward-looking concept and depends on the future income that can be generated from that asset. For example, if one spends a huge amount of money on a project that cannot provide any future income, then in principle this project is worthless. Meanwhile, if we were to value the water assets at their replacement cost, then the figures involved would be astronomical in view of today's huge construction costs. Historically, inflation in Hong Kong has been high; therefore, the replacement cost of assets would be several times more than their book value. If investors were required to pay the replacement cost for the water assets, they would demand a commensurate high return, and water bills would have to be increased tremendously. Hence, we suggest that the concept of "indicative value" be employed to price the water utility.

The Use of Indicative Value

The concept of indicative value has been used by regulators in the United Kingdom to value existing assets as if the regulatory regime had not changed (Cowan 1994). This concept implies that existing owners would neither gain nor lose from the change in regulatory regime. In order to apply the concept to the Hong Kong situation, we must calculate the water utility's financial value to the Hong Kong government, if the government continues holding the assets and to earn the same amount of cash flow. The following describes the method of arriving at an indicative value for the utility.

According to finance theory, the value of an asset depends on the expected future cash flow generated by the asset and the rate of return required by the investor or cost of capital in acquiring that asset. The fair value of an asset is the discounted value of the expected future returns to the holder of that asset. In order to value the water utility during the selling process, we have to carefully consider the determination of two factors: the expected future cash flow and the required rate of return.

Cost of Capital

Investors would finance a project only if the asset can provide sufficient returns as demanded. There are various methods available to determine the required rate of return, and the government has identified three that are particularly appropriate to the Hong Kong situation: the general approach, the economic approach, and the Capital Asset Pricing (CAP) approach (Finance Branch 1994).

General Approach, Economic Approach and CAP Approach

According to the general approach, the required rate of return should be set at a level that compensates investors for the effects of inflation as well as providing real growth for furthering the investment. The economic approach demands that return on the asset be equated with the return that the funds involved could have earned from an alternative investment. This is equivalent to the opportunity cost concept in economic theory. However, there are certain difficulties in the practical application of either approach. In the general approach, while the inflation rate can be calculated with little ambiguity, it is not an easy task to set a real growth rate with objectivity. In the Economic Approach, due to the unique nature of government utilities, there are difficulties in deciding what an appropriate alternative investment would be.

Recognizing the inherent problems of these two approaches, the government has suggested the use of the CAP approach in setting the target rates of return for various government utilities (Finance Branch 1994). The CAP approach is in fact considered an appropriate approach in the capital budgeting process, and it is widely used in business. We share the government's view regarding the use of the CAP approach to determine the target rates of return for public utilities. We also suggest that the same approach be used to assess the value of a government asset when it is to be sold to the public.

The CAP approach relies on the Capital Asset Pricing Model (CAPM) which was developed in the 1960s and has become popular for calculating required rates of return. The idea behind the CAPM is that investors are averse to risk and are only prepared to accept more risk if an investment can provide a correspondingly higher return. The risk of an asset has two components: one is unique to the asset itself and the other is related to overall market movements. For example, the fall in a company's share price may be due to an incorrect decision on the part of its management, or it may be due to a world-wide price slump. While investors can avoid the former risk by holding a well-diversified portfolio so that profitable and unprofitable stocks can balance one another, nothing can be done to eliminate the second type of risk. Investors therefore ask for compensation in order to bear this market risk (also called systematic risk). The higher the market risk of an investment, the higher the return investors will demand.

In the Capital Asset Pricing Model, the expected rate of return of an asset $E(R_i)$ depends on three factors: the rate that can be earned from a risk-free investment (risk-free rate, R_f), the return earned from a widely diversified market portfolio $E(R_m)$ above the risk-free rate (market risk premium) and the risk factor (beta, β) of the project itself. By risk factor we mean the magnitude of the correlation between the return on the asset itself and that of the market portfolio. This is exactly the risk that investors cannot avoid and should be compensated for. The expected rate of return is equal to the risk-free rate plus the market risk premium multiplied by the risk factor, the formula being $E(R_i) = R_f + \beta\ [E(R_m) - R_f]$.

The Cost of Capital Estimation

In other countries it is common to use the yield from government bonds to represent the risk-free rate. As Hong Kong does not have a well-established government bond market, we suggest that the six month interbank offer rate be used as a surrogate. The idea is that large investors can negotiate with banks to earn interest from deposits tied to the interbank rate. Such deposits have minimal risk, as has been evidenced over the past few years. With regard to the return from the diversified market portfolio, the rate of return from the All Ordinary Index compiled by the Stock Exchange of Hong Kong can serve as a suitable proxy. Some other investors would prefer to use the return of the Hang Seng Index as the market return proxy, but we believe that the All Ordinary Index is more appropriate, as it has a wider coverage and can better represent the whole market.

Since there is no similar listed company in Hong Kong engaged in the water industry, it is not easy to attribute a risk factor to the

Chapter 6

Table 6.1
Estimated Risk Factors of Electricity Companies in Hong Kong

Company	Leveraged Risk Factor (LRF)	Average Debt / Equity Ratio (D / E) during 1991–95	Unleveraged Risk Factor (URF)
CLP	0.92	0.253	0.76
HEC	0.98	0.174	0.86
Average			0.81

Note : The unleveraged risk factor (URF) is calculated by dividing the leveraged risk factor (LRF) by $1 + (D / E)(1 - t)$. The applied tax rate t is 16.5%.

privatized water utility. We suggest that the average risk factor of other related listed public utilities be used as the proxy of beta for our water company. The risk factors of the China Light and Power Company Limited (CLP) and the Hong Kong Electric Company Limited (HEC) are two suitable candidates. These two electricity companies have been selected because they share some common business features with the water utility.

The electricity companies and water utility are similar in that they all distribute their products to consumers through networks. They have to invest a huge amount of resources into building large-scale production plants and distribution networks, which have little resale value. Their products also exhibit similar income and price elasticity. Furthermore, the newly created water company will be facing some form of regulatory control in setting prices. This is similar to the regulatory risk encountered by the shareholders of the electricity companies. Last but not least, the water company has to rely on a long-term contract with China for its water supply, and this sort of supplier risk is shared by CLP in its purchasing of electricity from China's Daya Bay Nuclear Power Station.

Based on monthly data from 1991 to 1995, we have used regression (a statistical technique) to calculate the betas of the two electricity companies. We regressed the raw return (including dividend yield) of the two electricity stocks on the return of the All Ordinary Index and obtained the estimated beta values.

An estimated beta value represents the leveraged risk factor faced by equity holders, which in turn is affected by the leverage level (financial risk) of the company. For example, if the company is highly leveraged, then much of its earnings must be used in paying off the interest expenses, causing the residual return to equity holders to be subject to wide fluctuation. In order to determine the risk factor of the whole company net of financial risk (the company's unleveraged risk factor), we have to divide the leveraged risk factor by $1 + (D/E)(1 - t)$, where D is the company's market value of debt, E is the company's market value of equity and t is the applied tax rate. That is to say, the unleveraged risk factor (URF) and the leveraged risk factor (LRF) are related as follows:

$$URF = LRF / [1 + (D/E)(1 - t)]$$

We do not intend to go through the mathematical steps taken to derive the above relationship, but we would like to discuss the interpretation of this calculation. If a company does not have any debt outstanding, then its overall cash flow goes to the equity holders, and the two risk factors would be equal. If the company incurs a larger amount of debt, the risk faced by the equity holders will be higher than the risk of the whole company. Furthermore, the inclusion of a tax deduction from interest expenses will make a company's cash flow to equity holders less unstable, and therefore the unleveraged risk factor has to take into account the tax rate.

The estimated leveraged and unleveraged risk factors of the two electricity companies are presented in Table 6.1. Since both CLP and HEC also engage in some other non-electricity related businesses (e.g., property development) which are far more risky than their core businesses, the calculated risk factor somewhat overestimates the risk level of their core electricity business. Nevertheless, such overestimation should not be too significant, as the vast majority of the assets of the two companies are still used for generating electricity.

Chapter 6

Table 6.2
Operating Accounts of the Water Authority

	1995–96 Actual	1996–97 Estimate*	1997–98 Forecast*	1998–99 Forecast*
Revenue (HK$ million)				
Chargeable supplies	2,374.8	2,384.8	2,653.8	2,992.6
Contribution from rates	2,127.1	**	**	**
Contribution from gov. for free allowance to domestic consumers	384.3	404.9	428.7	490.9
Government uses	164.5	168.6	187.8	211.4
Fees / licences	41.3	44.7	49.2	53.9
Interest from deposits	45.9	44.6	49.5	56.0
Total Revenue	5,137.9	–	–	–
Expenditure (HK$ million)				
Staff cost	1,275.0	1,420.0	1,561.9	1,712.4
Bulk Purchase of water	1,541.2	1,788.9	2,027.8	2,290.4
General Operating expenses	1,221.9	1,291.4	1,332.1	1,405.2
Central administration	89.0	99.3	109.2	119.7
Depreciation	400.5	459.4	530.1	621.3
Total Expenditure	4,527.6	5,059.0	5,561.1	6,149.0
Taxation	0.6	–	–	–
Operating surplus	609.7	–	–	–
Average Net Fixed Assets (ANFA)	15,711	17,961	20,320	22,714

Sources: Figures for 1995–96 (except ANFA) are extracted from the Treasury, 1996, figures for other years (except ANFA) are extracted from the Water Authority, 1996, figures of ANFA are extracted from the Finance Branch, 1994.

Note: * The figures for 1996–99 were obtained based on the assumption that an annual increase in water charges of 9.9% for domestic supplies and 10.8% for non-domestic supplies is to have gone into effect as of January 1997 and that a 12.1% increase for both will go into effect in 1997–98.

** The figure of Contribution from Rates as presented in the document submitted to the Water Authority (1996) is based on the assumption that the contribution will just cover the operating deficits. As it did in past years, Contribution from Rates greatly exceeded the amount required to cover the deficits. We think the figures in the document may change substantially, and we therefore omit the figure here. It will not affect our analysis.

During the five-year period from 1991 to 1995, the average six month interbank offer rate (proxy for risk-free rate) equals 5.17%, the average compound return on the All Ordinary Index (proxy for market return) equals 22.5%, and given that the estimated average unleveraged risk factor of the electricity companies has a value of 0.81, the nominal rate of return of the water utility as required by private investors is calculated as :

5.17% + 0.81 x (22.5% − 5.17%) = 19.2%

Given that the average inflation rate in Hong Kong was 9.2% during the same period, the real required rate of return is 10% (19.2% − 9.2%). These figures will be used in the subsequent analysis.

Expected Future Net Cash Flow

Based on a discussion paper submitted to the Legislative Council about the 1996–97 water tariff revision proposal, the government has made a projection for the financial performance of the waterworks operation up to the fiscal year 2000–01. Part of the government's projection is extracted and shown in Table 6.2. We also present the actual financial performance of the water authority during the fiscal years ending on 31 March from 1993 to 1996 in Table 6.3.

It is evident from Table 6.3 that the actual return on the Waterworks Average Net Fixed Assets (ANFA) averages 5.7% over the four-year period from 1993 to 1996. This is somewhat below the target rate of return (ranging from 6.5% to 8% in different years) set by the government. The inability of the Water Supplies Department to achieve the existing target rate of return is due to the rapid expansion of ANFA (anticipated at an average of $3 billion per annum) and the reduction in actual sales during the period.

Now, suppose the government adopts the proposal to privatize the water authority by selling its assets to the public and that such a sale is scheduled to take place during the fiscal year 1998–99. According to the required rate of return calculated above, if all the

74 Chapter 6

Table 6.3
Financial Performance of the Water Authority
(HK$ million)

	1992–93	1993–94	1994–95	1995–96
Revenue	3,990.0	4,347.2	4,785.9	5,137.9
Expenditure	3,114.8	3,480.1	3,867.6	4,527.6
Operating surplus before taxation	875.2	867.1	918.3	610.3
Taxation	153.2	54.7	58.3	0.6
Operating surplus after taxation	722.0	812.4	860.0	609.7
Average Net Fixed Assets (ANFA)	11,443.6	12,706.9	14,099.9	15,711.0
Actual return	722.0	812.4	860.0	609.7
Target return	801.1	876.8	1,128.0	1,021.2
Actual return as % of ANFA	6.3%	6.4%	6.1%	3.9%
Target return as % of ANFA	7.0%	6.9%	8.0%	6.5%

Source: Treasury, *Operating Accounts of Hongkong Government Utilities, 1993–96.*

ANFA are sold to the private sector at book value (HK$22,714 million) during the fiscal year 1998–99, then investors will ask for an annual surplus of HK$4,361 million (HK$22,714 million x 0.192) from the waterworks operation in order to make the purchase of the privatized water utility's shares attractive. According to the past and projected performance of the water authority as presented in Table 6.2 and 6.3, such an annual surplus is deemed to be unachievable unless there is a tremendous increase in revenue. For example, using the projected total expenditure of HK$6,149 million during 1998–99, revenue has to increase to HK$10,510 million (HK$6,149 million + HK$4,361 million), which is more than double 1995–96 revenue.

As an alternative, we suggest that the surplus figure be equal to the target return as set by the government in the absence of the privatization programme. Thus, the surplus figure will equal HK$1,476.4 million (HK$22,714 million x 6.5% where 6.5% is the target rate of return in 1996). This will prevent any price increase brought on by the privatization process per se. This means that we project the surplus that the water utility would earn on existing assets if the sale had not taken place. Since in Chapter 5 we suggest that the water utility be subject to the price-cap regulation in determining the charge increase after privatization, the subsequent revenue of the company will be indexed to reflect inflation. We therefore have to use the real required rate of return (i.e., 10%) as the discount factor. Discounting the annual surplus of HK$1,476.4 million by the discount rate of 10% results in the market value of the privatized water utility, which is equal to HK$14,764 million. This is the fair value of the existing assets at the time of sale, without taking any further investment obligation into consideration.

Since the book value of the waterworks assets in 1998–99 (HK$22,714 million) greatly exceeds the proceeds from sale, people may ask whether the government has incurred a huge loss by selling the assets at such a discount to the private sector. The answer is both yes and no. Were the water utility still in government hands during 1998–99 and were it restricted to earning the 6.5% target rate of return, the government would be losing 12.7%, as the funds it injects could be earning 19.2% if invested in other business projects with similar risk. This means that the government is losing in an economic sense (by subsidizing water users) as long as it holds the water utility. The 12.7% is an annual loss during 1998–99, and similar losses will be incurred year after year if the government continues to set a target rate lower than the required rate of return of assets with similar risks. Selling assets at a discount, which is a capitalized loss, is just the other side of the same coin. In other words, the market value of a company has to be lower than its book value if the assets cannot provide a return that satisfies investors; such cases are not rarely observed in the stock market. Actually, the ratio of book to market value of an asset can be shown to be equal to the ratio between the cost of capital and the actual rate of return. The higher the cost of capital above the actual rate of return, the lower the market value when compared to the book value.

We can also analyse the price impact on consumers under our privatization proposal. As has been discussed in Chapter 5, we want

Table 6.4
Projected Financial Figures for the Water Authority, 1998–99
(HK$ million)

Average Net Fixed Assets (ANFA)	22,714
Target Rate of Return on ANFA	6.5%
Required Surplus (22,714 x 0.065)	1,476.4
Plus	
Staff cost*	1,369.9
Bulk purchase of water	2,290.4
General operating expenses*	1,124.2
Central administration*	95.8
Depreciation	621.3
Subtotal	5,501.6
Minus	
Contribution from Government for allowance to domestic consumers	490.9
Government uses	211.4
Fees / licences	53.9
Interest from deposits	56.0
Subtotal	812.2
Required Chargeable Supplies	6,165.8

Source: Water Authority 1996.
Note: * These figures were obtained based on the assumption that such costs can be reduced by 20 per cent as projected by the government due to the privatization programme.

to assume the achievement of productivity enhancement in the Water Supplies Department before privatization. Suppose that there is a 20% cut in staff and operating expenses due to re-engineering. Adding the new expenditure to the required surplus figure and subtracting income from other sources, the required revenue from chargeable supplies from consumers has to be equal to HK$6,165.8 million during 1998–99 (See Table 6.4). Compared with the government's initial projected revenue from chargeable supplies (HK$2,992.6 million), this represents a 106% increase, and therefore the tariff on water supplies has to be adjusted accordingly. However, we should be reminded that this figure is calculated based on the assumption that there is no longer any contribution from rates to subsidize the waterworks operation. As consumers have to pay more for using water, at the same time they are also benefited by paying lower rates. We think such a move is feasible and that it could achieve a better allocation of resources.

Concluding Remarks

After privatization, the water utility will be a publicly listed company on the stock exchange. The company will be able to finance its further expansion through the market. Being a listed company will also enable the public to share in the profits earned by this public utility. The government can retain part ownership of the company during the initial stage of privatization. Such a connection between the government and the company will ensure a smooth transition of the public utility into the private sector. However, the privatized company should be run on pure commercial principles and should be accountable to its shareholders. It is these commercial principles that will drive the company to achieve flexibility and efficiency, which are the ultimate goals of privatization.

A priority programme of share subscription for the employees of the company can be included as part of the privatization process. This priority in share subscription can provide compensation to employees for the transformation of working status, and it can provide an inducement to work efficiently since they will be sharing part of the company's profit. Their monetary reward is thus tied to the long-term performance of the company. Furthermore, the U.K. experience shows that a wider distribution of share ownership will reduce customer opposition to privatization.

The water utility will have to make new capital investment after privatization. Funding can come from either borrowing or from equity financing but no longer from government tax revenue. Water charges will have to be increased to finance the acquisition of new assets, and investors should be allowed to earn a return sufficient to cover their capital cost. But the annual price increase will be governed by the price-cap formula described in Chapter 5. The K factor in that formula will take into account the price increases required to finance further expansion. People should not worry unduly about any tremendous water charge increases after privatization. Nevertheless, since there will not be any government subsidy, the rate of charge increase will be somewhat higher than what we are now facing. Customers should be consulted in determining the level of investment.

If the privatized water utility is to be allowed to invest in new assets to improve water services, then users have to be prepared to face higher charge increases. This is unavoidable if the "users pay" principle is adopted and if the water company is to be run on commercial principles. However, both the price-cap regulation and the take-over threat mechanism in the stock market will put pressure on management to improve productivity. Customers will enjoy benefits resulting from productivity enhancement, and taxpayers will no longer be required to subsidize water services. Altogether, we stand to gain from the privatization of the water utility.

CHAPTER 7

Privatization of the Drainage Services Department

The Nature of Water Pollution

In Hong Kong, the government is responsible for all the capital costs incurred in building the infrastructure for sewage and flood control. The Sewage Service Trading Fund was designed to recoup the operation and maintenance costs only. Unlike a target rate of return imposed on the assets used in supplying water, the rate of return on the assets used in sewage disposal and flood control was set at zero. A huge fiscal reserve accumulated over the past decade had enabled the government to bear the capital costs incurred in improving water quality. In addition, there were other economic arguments for public provision of sewage and flood control systems. Still, how long can this free ride last?

The Problem of Externality in Controlling Water Pollution and Flooding

In situations where the market fails to achieve the optimal allocation of resources, that is in cases of "market failure" where government intervention is required to correct the situation. The problem of pollution is often cited as a case for government intervention. Also, in textbook economics, private cost can be distinguished from social cost. Private cost measures the value of the highest-valued alternative uses of the resources available to the producer (the person who takes the action); social cost measures the value of the highest-valued alternative uses of the resources available to the whole society. In some situations, a person may only consider his or her private cost, rather than the social cost, of an action. An external cost, or a harmful "externality" exists when one person's action imposes a cost on others but that the person bears no cost. In such a case we say that there is a divergence between private and social costs. The divergence between private and social costs is termed an "external cost". Water pollution is a typical example of such an externality. When a person pollutes the society's water resource, his action will impose a harmful effect on other people. Apart from hurting, say, the fishermen, the damage may reduce the value of property owned by residents living near the polluted watercourses.

Traditionally, whenever there is a divergence between private and social costs, government intervention is called for to tax the firm generating the external cost. This would raise its marginal cost of production and would thus eliminate the excess output. Following this logic, economists for many years subscribed to the belief that certain penalties or restrictions should be imposed on those responsible for harmful externalities. Such measures could range from taxes or fines relative to the amount of damage caused, to the physical removal of the firm causing the externality, and to an outright prohibition on production. All these remedies would necessitate an expanded government role.

This view was challenged in 1960, however, by Ronald Coase. In Coase's paper (1960), he re-analyzes the problem of social cost from the ground up. Coase argues that the objective in using scarce resources is not to minimize damage but to maximize total gains. Externalities are, in general, part of the result of any useful and productive activity. It would be wasteful to eliminate the productive activity simply to avoid small damages. A sensible objective is to achieve the right balance of damages and gains, or the right balance of costs and benefits. According to Coase, if property rights are clearly defined and transaction costs are zero, it is possible to achieve the right balance of damages and gains by contract, whatever the legal liability may be.

Coase's argument is that the existence of an external effect, be it harmful or beneficial, indicates that there are potential gains from the trade involved. These gains can be captured through contracts between those responsible for the effect and those affected by it. The initial assignment of property rights would not affect resource allocation; it would only affect income distribution. The party with the right to use the resource would exclusively enjoy the income derived from it. The person who is liable for the damage has to compensate the other party who suffers from his action. Unless the gain from the damage exceeds the amount of compensation, there will be no market transaction. Hence, optimal resource allocation can still be achieved through market transaction of the "externality". Coase notes too that the presence of high transaction costs may limit the possibility of market transactions. If transaction costs saved are greater than the gains from exchange, then there is no inefficiency if an "externality" exists.

In the case of sewage disposal and flood control services, it is difficult to clearly delineate legal liability and then allow the market to work. The use of private contracts to identify polluters and measure indemnity can be costly. In addition, the construction of private sewage and flood control systems may often involve many owners of private land, and this would increase transaction costs in the negotiation process over connections and interconnections. The high transaction costs incurred in private contracting can therefore make a case for the public provision of sewage and flood control systems.

The Public Good Nature of Sewage Disposal and Flood Control

In economics, a good is a private good if its consumption by any one person would reduce the amount available for others. For example, if you use a piece of land to build a house, you deny someone else the opportunity to cross it, to park a car on it or to build another house there. On the other hand, a good is a public good if its consumption by any one person does not reduce the amount available for others.

Figure 7.1
Marginal Cost Pricing under Natural Monopoly

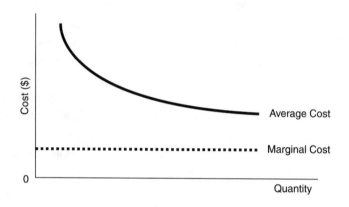

In other words, a public good is one which can be consumed concurrently by many individuals. Once a sewage system or a flood control system has been installed, the benefit is available to all people. Everyone can enjoy a cleaner environment and can benefit from avoiding the damage caused by flooding.

A public good is in joint supply in the sense that once it is produced, any given unit of the good can be made equally available to all. We cannot divide the good among different individuals. For a private firm producing a public good to charge a price for its use would restrict its use and reduce its total use value. Since the public good has already been produced, reducing its use would constitute a waste equal to the loss in the total use value derived from the good. But if no price is charged, how will funds be collected to pay for its installation? It is then argued that public goods should be provided by the government and financed by taxes.

The Natural Monopoly Status of Sewage Disposal and Flood Control

When an industry enjoys substantially increasing returns to scale, meaning that when the average cost of production declines continuously, we say that the industry is in the condition of a natural monopoly. A single firm in the industry can provide the good at a lower total cost than several smaller firms. It has been argued that substantial economies of scale are found in sewage treatment and flood control. A large treatment plant costs much less on average than several smaller plants. Sewage disposal and flood control are examples of natural monopoly.

In a natural monopoly, the marginal cost is always less than the average cost. As was mentioned earlier, optimal resource allocation requires a good or service to be sold at a price equal to its marginal cost of production. Requiring a private firm to produce at the level of output where the price is equal to marginal cost would mean that the firm would suffer from chronic loss, since the average cost is higher than the price. A private firm based on marginal cost pricing could not exist. If the price charged is based on average cost (which includes capital costs), then the output level will be less than optimal (see Figure 7.1). It is argued, therefore, that the government should intervene. If the government bears the capital costs and charges users according to the marginal operating cost, the amount of output will not be restricted to below the optimal level.

To conclude, the externality effect, the public good nature and the natural monopoly status in the provision of drainage services can serve to support the existing charging system in Hong Kong. In other countries, a long-run marginal cost pricing system is often applied to cover the capital costs of waste water treatment, but we recommend that the Hong Kong government continue its policy of financing the capital works from general revenue. The "polluter pays" principle should only be applied to recovering operating costs in waste water treatment. So long as sewage charges continue to cover only operating costs (but not capital costs), the current system will perpetuate the cross subsidy between taxpayers and polluters; it can be expected to lead to an excessive level of pollution. However, this charging system is more politically feasible, and the subsidy can also be justified because of the natural monopoly and public good nature of sewage disposal.

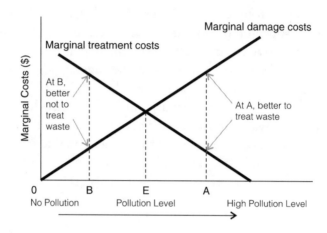

Figure 7.2
The Optimal Level of Water Pollution

Source: Scott (1995), p.148.

The Optimal Level of Water Pollution

We have found pitfalls in discussions of water pollution problems. The first common pitfall is the assumption that it is the responsibility of the government to pay all the costs of waste water treatment. In order for it to do so, there must be sources for financing the expenditure. Before 1995 the cost of sewage collection and treatment came entirely from general revenue. The public was not required to pay for the cost of treating the waste water it discharged. As a result, the public was unable to see the cost of sewage disposal, and there was no incentive for the public to reduce water pollution.

Another common pitfall concerns the optimal level of water pollution. Ardent environmentalists often criticize the deteriorating level of water quality in Hong Kong. They frequently argue that the cleaner the water, the better living standards will be. They tend to overlook that clean water is not a free good. There is an opportunity cost, in terms of output forgone, involved in reducing water pollution. There are also treatment costs involved in cleaning the water. The optimal level of water pollution is achieved when the marginal damage cost of water pollution is equal to the marginal treatment cost.

Figure 7.2 shows the marginal damage costs and marginal treatment costs of water pollution. At pollution level A, the marginal costs of damage to society (in terms of output forgone, loss in health and loss in scenic value) is much higher than the marginal treatment costs. Society would be better off with less water pollution. The benefits derived from waste water treatment can easily outweigh the costs. On the other hand, at pollution level B, the treatment costs outweigh the damage costs at the margin. It does not pay to reduce water pollution to such a low level. The optimal level of water pollution, therefore, is at pollution level E, where the marginal damage costs are equal to the marginal treatment costs.

In order to induce polluters to reduce pollution to this optimal level, they should be required to bear the damage costs. The government could charge a pollution tax equal to this amount. Alternatively, it could issue licences for pollution up to the level E and could use a bidding system to allocate the pollution rights. Although these two measures could help achieve the optimal level of pollution, they would not encourage efficient treatment of waste water. As damage costs or treatment technology will change over time, the optimal level of pollution will also change. Under the pollution tax or licensing arrangement, once the static level has been reached, there is no further ongoing incentive to reduce water pollution (Scott 1995).

In Hong Kong, the Drainage Services Department exempts residents or business operators from paying sewage charges if they have their own waste water treatment installations. The department also reduces the sewage charges if it is satisfied with the treatment work done by individual firms. This incentive scheme has encouraged people to minimize damage and treatment costs. However, as the government has been responsible for the capital costs of sewage treatment, and large treatment plants have enjoyed substantial economies of scale, public and private sewage treatment companies are not on a level playing field. Moreover, as the

Drainage Services Department is a public enterprise that does not enjoy price and wage flexibility, market pressure on cost effectiveness is limited. In view of these incentive problems inherent in the existing water treatment market, we recommend the privatization of the Drainage Services Department. Since we have argued for government provision of the capital works of sewage treatment, the privatization arrangement would differ from that for the Water Supplies Department. A system of contracting out would be adopted to increase the efficiency of sewage disposal in Hong Kong.

Contracting Out Government Businesses

Contracting out is the process whereby a government hires under contract a private firm to perform, over a defined period of time, some specific service that might otherwise be provided by public employees using government equipment and facilities (Utt 1991). Private companies are invited to submit bids to perform services according to government specifications. If the bids submitted by private operators are below the cost of the services as carried out by the government, then public expenditure can be reduced if the private firm is invited to provide the service.

The Benefits of Contracting Out

There are various forms of benefits associated with contracting out. The first one is the improvement in efficiency. According to economic theory, private firms have a clearly defined goal of profit maximization and are thus likely to perform better than public bodies, whose goals are diffused and uncertain. Moreover, public choice economists hold the view that politicians and bureaucrats pursue their private interests rather than the will of the citizens. Government officials tend to formulate policies so as to maximize their arena of power and engage in such behaviour as risk aversion and non-optimal pricing, employment and investment. If operations are privatized, the profit motive will put pressure on cost control and bring about increased efficiency, and eventually these advantages will be enjoyed by consumers if competition takes place among private operators.

The second benefit of contracting out is flexibility. Private operators are likely to be more flexible, as they are not bound by civil service rules and regulations in hiring or disciplining employees. They will not be caught up in so much red tape as the government departments.

Thirdly, government enterprises are often monopolies in their field of business because they face no threat of competition. If an operation is contracted out and allocated to different parties, then there will be some form of competition introduced into the market. Although the private operators may not be in direct competition for customers due to geographical or functional reasons, people can still make comparisons among their respective service quality or charge levels and consequently the operators are under pressure to improve.

The government has to address some major issues in the process of contracting out government businesses. The first is the identification of the optimal contract length. An overly long contract will bring inflexibility to both parties, while a short one will make it difficult for the operator to demonstrate its ability. The second issue concerns contract specification. Contracting out is more suitable for those goods or services for which the characteristics of supply can be specified with precision. Specifying precisely the terms and conditions of service quality will make it easier to assess the performance of the operators. Finally, it must be recognized that contracting out still requires some regulation because contracts have to be enforced and contractor performance monitored.

The Experience of Contracting Out in Hong Kong

The concept of contracting out government properties for private management is not new to the Hong Kong community. Nowadays, we have certain public utilities that are under government

Table 7.1

Time Table of Privatizing Management of Government Tunnels

Tunnel	Date of Commencement of Operation	Date of Commencement of Management Privatization	Name of Management Company
Lion Rock	November 1967	January 1993	Serco Guardian Limited
Aberdeen	March 1982	September 1991	Cross Harbour Tunnel Co. Ltd.
Airport	June 1982	January 1993	Serco Guardian Limited
Shing Mun	April 1990	January 1993	Mack and Co Tunnel Management Ltd.
Tseung Kwan O	November 1990	January 1993	Mack and Co Tunnel Management Ltd.

Source: Hong Kong Government, *Annual Report.*

ownership but managed by private companies on a commercial basis. Examples of these include abattoirs, road tunnels and car parks, among others. The following presents a brief description of current practice in two of these cases.

Road Tunnels

Hong Kong has a total of eight road tunnels for vehicles. Five of these are government owned, while the other three are owned by private companies. The five government tunnels, in order of commencement of operation, are the Lion Rock Tunnel, the Aberdeen Tunnel, the Airport Tunnel, the Shing Mun Tunnel and the Tseung Kwan O Tunnel. Except in the case of the Airport Tunnel, drivers have to pay tolls for using these tunnels. All of them were initially managed by the government, and the shift to private operation has been gradual.

The idea of contracting out the government tunnels dates back to October 1988, when the government started to think about the best proposal for transferring the tunnels to non-government bodies. The government has considered various possibilities, such as setting up a separate subsidiary to take over the ownership and management of the tunnels, the divestiture of tunnels to the private sector, and putting out the management of the tunnels to private tender. The responsible authority finally adopted the third alternative — for the first proposal could do little to lessen the burden on the government while the second would involve other difficult issues like the valuation of assets and the setting up of regulatory bodies to monitor and control service quality. After the method of contracting out had been decided upon, the Aberdeen Tunnel was selected as the pilot project to be transferred to private management in September 1991.

The Cross Harbour Tunnel Company Limited was chosen as the management company for the Aberdeen Tunnel upon its privatization. Approximately 400 civil servants were affected by this new scheme. About 60 of the them remained in the new management company; the rest were either transferred to other government departments at the same salary level, chose early retirement, or simply lost their jobs.

Following the success of the management privatization of the Aberdeen Tunnel, other government tunnels followed suit. Table 7.1 shows the starting dates of the privatization of these tunnels.

After privatization, the ultimate ownership and control of the tunnels remained in the hands of the government, but the daily operations were passed on to the private tunnel operators. To ensure that the service quality of the latter is up to standard, the government asks them to submit reports, and government inspectors regularly pay unexpected visits to the tunnels. The operators are entitled to receive a management fee in return for their provision of services. The level of the management fee is proportional to the total amount of the toll fares collected. All toll increases must be approved by the Legislative Council.

Table 7.2 presents a comparison of the performance of government toll tunnels before and after their operations were contracted out. For ease of comparison, we divide the table into two panels. The first panel covers the financial years 1985–86 to 1989–90, when only two tunnels (Lion Rock and Aberdeen) were in use, and the second panel covers the financial years 1991–92 to

Chapter 7

Table 7.2
Operating Accounts of Government Tunnels
(A) 1985–86 to 1989–90 (Tunnels in use: Lion Rock and Aberdeen)
(HK$ million)

	1985–86	1986–87	1987–88	1988–89	1989–90
Total revenue	115.4	133.7	150.1	161.8	167.0
Expenditures					
Staff costs	21.9	23.4	25.7	28.2	33.2
Running expenses	24.8	36.5	30.0	27.3	37.1
Subtotal	46.7	59.9	55.7	55.5	70.3
Depreciation	19.2	19.4	19.4	19.3	19.6
Total expenditure	65.9	79.3	75.1	74.8	89.9
No. of vehicles using tunnels (million)	38.3	44.4	49.9	53.7	55.4

(B) 1991–92 to 1995–96
(Tunnels in use: Lion Rock, Aberdeen, Shing Mun and Tseung Kwan O)
(HK$ million)

	1991–92	1992–93	1993–94	1994–95	1995–96
Total revenue	349.7	367.4	388.0	410.3	441.8
Expenditures					
Staff costs	69.0	57.7	4.9	5.0	5.3
Running expenses	71.0	58.4	19.5	20.0	19.8
Management fee	11.9	39.5	107.3	118.4	128.9
Subtotal	151.9	155.6	131.7	143.4	154
Depreciation	75.9	70.7	72.0	72.7	72.6
Total expenditure	227.8	226.3	203.7	216.1	226.6
No. of vehicles using tunnels (million)	65.4	68.7	74.6	80.3	86.8

Source: Treasury, *Operating Accounts of Hongkong Government Utilities, 1981–96.*

1995–96, when all four toll tunnels were put into operation within the financial year.

During the five-year period between 1985–86 and 1989–90, the number of vehicles using tunnels increased at an average annual rate of 9.7%. At the same time, the operating expenses excluding depreciation increased at an average rate of 10.8% annually. Since operating expenses were increasing faster than output level, unit operating expenses went up slightly. After the government had contracted out the Aberdeen Tunnel to private companies in 1991, the total expenditures from tunnel operation began to feature a "management fee" item. The privatization of Lion Rock Tunnel, Shing Mun Tunnel and Tseung Kwan O Tunnel in January 1993 caused the management fee item to jump to HK$107.3 million during the fiscal year ending in 1994, while at the same time the staff and running expenses incurred by the government were cut to lower levels. It is worthwhile noting that total operating expenditures (including the management fee but excluding depreciation) dropped from HK$155.6 million during the fiscal year 1992–93 to HK$131.7 million during the fiscal year 1993–94. This represented a one-step 15% gross cut in operating expenditures. Further, the net cut in expenditures due to contracting out should be even higher if we also take into account the inflation factor.

In the absence of deterioration of service quality, contracting out the toll tunnel operation has thus shown that private companies are more efficient than the government in utilizing resources. Moreover, it has also been found that unit operating expenses increased less rapidly after the contracting out. From 1993–94 to 1995–96 operating expenses before depreciation increased at an average annual rate of 8.1%, while the total number of vehicles using tunnels was increasing at the rate of 7.9%. Compared with the figures for the 1985–86 to 1989–90 period, the unit operating expenses were lower after privatization.

Table 7.3
Cost Structure of Sewage Services Operation, 1995–96
(HK$ thousand)

Staff costs	314,322 (49.2%)
Light and power	65,878 (10.3%)
Hire of services and professional fees	35,799 (5.6%)
Operation and maintenance expenses	130,270 (20.4%)
Rental and management charges	22,008 (3.4%)
General operating expenses	57,633 (9.0%)
Provision for doubtful debts	9,500 (1.5%)
Depreciation	2,783 (0.4%)
Audit fees	800 (0.1%)
Total Costs	638,993 (100%)

Source: Sewage Services Trading Fund, *Annual Report*, 1995–96

In October 1994 the Treasury suggested that the target rate of return on the toll tunnels' Average Net Fixed Assets (ANFA) should be set at 13.5% (Finance Branch 1994). The Shing Mun and Tseung Kwan O Tunnel tolls will thus have to be increased in order to meet this target, whereas the other two toll tunnels can enjoy a toll freeze for being above or in line with the target rate.

Car Parks

The government started to contract out the management of car parks to a private company in May 1984. In that year the Transport Department, after a tendering exercise, signed an agreement with Wilson Parking (Hong Kong) Pty Limited for the operation of ten multi-storey car parks for a three-year period. Wilson Parking successfully gained contract renewal after the expiration of the first contract and continued to be the single car parking management company appointed by the government for nine years. In 1993 the Transport Department adopted the Consumer Council's advice and decided to split the management contract between two separate companies in order to enhance competition and stimulate better service. As a result, Wilson Parking continued to operate seven multi-storey car parks, while the management of the remaining seven was won by a subsidiary of Wharf Holdings.

Under the existing management contract, the government retains the authority to approve the level of parking fees. The operators can keep a certain percentage of total revenue from car parking and provision of advertising spaces outside the car park buildings in return for their management effort.

Contracting Out Sewerage Operations

As we have earlier argued for economic efficiency reasons, the government should bear the capital costs of sewage treatment plants and equipment but make the polluters bear the costs of daily operation and maintenance of the sewerage system. It follows that the Drainage Services Department should retain the responsibility for designing and constructing the sewerage, sewage treatment and disposal, and storm water systems throughout the territory. At present, the operation and maintenance costs of sewage services are recovered by the Sewage Services Trading Fund. Workers involved with sewage work are government employees. This induces inflexibility in arranging the recruiting, hiring or disciplining of employees when compared with these same activities as conducted by private organizations. If sewage operations are privatized, much of the existing red tape can be minimized. The Drainage Services Department has to be restructured by contracting out the provision of sewage services to private sector companies, the government retaining its monitoring role and the responsibility for investing in fixed assets.

Table 7.3 presents the cost structure of sewage service operation during the fiscal year 1996. The figures show that the majority of expenditures arose from staff costs (49%), while general operating expenses and operation and maintenance expenses accounted for 40%, and light and power made up the remaining 11%. The cost structure of sewage services operation is similar to that of toll tunnel operation before privatization, although the nature of the two businesses are different. The staff cost of both account for about half of their respective total expenditures. Since the government toll tunnels have been able to successfully cut costs after being taken over by private management, we predict that sewage services will also enjoy a cost reduction under the commercial operation principle.

We recommend the government to grant the franchise for providing sewage services via a competitive bidding process in which a bid would take the form of the proposed price to be charged for provision of services. The prospective firm that offers the lowest bid would be awarded the franchise. As long as there is sufficient competition at the bidding stage, the bidding price will be very close to the cost of the most efficient operating firm, and therefore the cost advantage can be transferred to the public.

The experience of contracting out the management of the government tunnels can be applied to the privatization of sewage operations. It is not a good idea to contract out sewage operations to private companies all at once. Just as it did in the case of the government tunnels, the government should first designate a pilot project for private management. After observing the performance of the private operator and assessing the problems associated with private operation, the government can then go ahead with the complete privatization of sewage operations.

As of 31 March 1996, the Sewage Services Trading Fund had a total establishment equivalent of 1,393 posts, of which 1,003 were deployed on the operation and maintenance of sewerage systems and sewage treatment facilities, 160 were on the management of the charging scheme and general administration of the Trading Fund, and the remaining 230 being on the design and construction of capital projects. (Sewage Services Trading Fund Annual Report, 1995–96). If in the future the government continues to bear the responsibility for construction work, while the daily operation of sewerage system is contracted out, then approximately 1,160 staff members in the Trading Fund will be affected. According to the experience of the privatization of government toll tunnels, some of these staff members would be shifted to the new private operator, while others would be transferred to other government departments or find other jobs.

Stages of Contracting Out

According to the plan of the Drainage Service Department, the Strategic Sewage Disposal Scheme (SSDS) Stage 1 is scheduled for completion by mid 1997 (see Figure 2.4 in Chapter 2). This scheme aims to collect all the sewage from Northeast Hong Kong Island, Kowloon, Tsuen Wan, Kwai Chung and Tseung Kwan O into a deep tunnel intercepting sewer system that will discharge the sewage through a long outfall into the Western Harbour after enhanced chemical treatment on Stonecutters Island. The Stage 1 programme will be followed by the SSDS Stage 2, in which an underground tunnel will be built to discharge the after-treatment waste water from Stonecutters Island further into the southern part of the Hong Kong water district.

In view of the completion schedule of these various sewage systems, we suggest that the Drainage Service Department should continue to operate the facilities under the High Priority Programme (which comprises the implementation of SSDS Stage 1 and six related Sewerage Master Plan projects). But at the same time, the department should consider the possibility of contracting out the operation of programmes under HPP and SSDS Stage 2 together to private operators after the completion of SSDS Stage 2. Since some of the existing schemes (e.g., the North-West Kowloon Sewage Treatment and Disposal Scheme) will be integrated into SSDS Stage 1 in mid 1997, the first two stages of SSDS will involve staff members from current government establishments as well as new staff members recruited due to the inception of the new programme. If the government decides to privatize the sewage operations of the HPP and SSDS Stage 2 in the future, then the newly recruited staff members under SSDS should not be accorded the usual government employment terms. In their employment contracts, these newly recruited staff members should be reminded of the possibility of being transferred to new private operators. As was mentioned above, existing staff members affected by the privatization process may be laid off, transferred to other government departments or employed by the new private management companies.

The third stage of the SSDS concerns the collection of the after-treatment waste water from northern Hong Kong Island into a discharge system; the fourth stage involves similar work in the southwestern part of Hong Kong Island. The government could consider contracting out the operation of the third and fourth stages of the SSDS to a private body if the privatization of the first and second stages is successful. After the completion of the four stages of the SSDS, around 75% of the sewage and waste water in Hong Kong will be treated before disposal. The remaining 25% will have to be discharged into Hong Kong waters after secondary treatment at the six large-scale treatment plants in Sha Tin, Tai Po, Yuen Long, Shek Wu Hui, Sai Kung and Stanley. There is no reason why the operation of these treatment plants should not be contracted out if we accept the principle of privatization. Eventually, all sewage treatment plants and systems throughout Hong Kong will be run by private organizations.

Measures to Reduce Governance Costs

It is recommended that the government should contract out the various sewage operations to more than one private operator. However, the number of contractors should not be too large, in order to allow operators to take advantage of economies of scale in the provision of services. For example, the operation of the HPP and SSDS Stage 2 might be allocated to one party, Stage 3 and Stage 4 to another party, and the other six treatment plants to other parties. The assignment of different projects to various parties is important for the success of the privatization programme and has several advantages. The services provided by different operators can be compared and the relative performance level of different operators can be assessed. The underperformers will face the threat of replacement by a competitor in the next contract.

Work done by several economists (e.g., Williamson 1976) has provided us with certain clues about the bidding process. When formulating the bidding process, the government should notice that a once-and-for-all bidding scheme is not feasible because it requires each potential bidder to specify the terms at which he is prepared to supply service now and the conditional terms at which he will supply service in the future. Such a complete contingent claims contract is impossibly complex to write, negotiate and enforce. For recurrent short-term contracts, there are also severe transactional problems at the initial award stage, at the execution stage, and at the contract renewal stage. The winning bidders may behave opportunistically by lowering the quality of service, which is difficult to define clearly in the bidding process. Penalty clauses in the contract should be designed carefully in order to eliminate this kind of behaviour.

When contracts are renegotiated, or when assets are transferred from existing operators to successor firms, asset specificity of human and non-human capital may cause certain problems. Employees of existing firms who have invested in a lot of training may find their services useless to successor firms. Moreover, since the physical assets of sewage operations are financed and owned by the government, while daily operations are carried out by private operators, serious principal–agent problems may arise. Private operators may overuse and pay little attention to the maintenance of machines that they do not own. Thus, contracts should stipulate the degree of effort required on the part of private operators to keep the facilities in good condition. The assessed value of the assets after contract expiration should not fall below expected levels after adjusting for depreciation due to normal usage. In order to mitigate all these contractual difficulties, the government has to introduce an extensive regulatory and arbitration apparatus. This should include methods of assessing plant and equipment values, routine auditing procedures, and arbitration processes to deal with disputes between existing and successor firms.

Finally, the Drainage Services Department should continue to take on the responsibility for planning and constructing the drainage systems and other drainage-related activities such as flood control. The Sewage Service Trading Fund would remain in existence after privatization — according to news in September 1997, it may dissolve in 1998. There would not be any change in the policy of allowing the fund to earn zero return on the department's physical assets. The task of the fund would be to collect sewage charges from polluters and channel the charges into the payment of the management fee to the private operators. The charges for sewage services would be tied to the bids submitted by the successful bidders. It is expected that the efficiency gained from actual and potential competition among private operators would help reduce the pressure for charge increases and help lessen the burden on consumers.

CHAPTER 8

Conclusion and Recommendations

Privatizing the Water Supply and Sewerage Industries

A major conclusion of this study is that revenue from water charges is insufficient to provide a reasonable return on the capital invested in the water industry, and that the government's Trading Fund will not solve the two basic problems of the water industry: rapid cost increases and low productivity. The "polluter pays" charging scheme through a trading fund arrangement, though being on the right track, will not work. As some analysts argue, the crux of the matter lies with privatization because without it there will be insufficient incentive for water supply and sewerage monopolists to contain cost increases. Successful attempts in Western countries have demonstrated that privatized water companies have effectively freed government from financing capital works through general taxation.

As a result of Hong Kong's democratization process, price increases by public utilities have been more and more politicized. Greater competition faced by the public utilities and low prices of services have caused share prices of public utilities to fall behind other stocks. Utility companies must therefore improve profit prospects by increasing productivity and lowering costs. Because of government subsidy, the performance of the water supply and sewerage industries compares unfavourably with private utility industries. Since the present analysis finds no sound economic reason for not allowing the private sector to provide water and sewage services, the government should consider the option of privatizing the water supply and sewerage industries.

Recommendations

Industry and Regulatory Structures

We recommend that the existing structure of vertical separation between the water supply and the sewerage departments be maintained. We recommend that the Hong Kong government take two levels of action in respect of drainage services. First, the present policy of financing the capital works from general revenue should be continued; and second, the "polluter pays" principle should continue to be applied to recover the operating costs in waste water treatment. The first level is justifiable because of the externality, the public good nature and the natural monopoly status in the provision of drainage services. As to water supply, however, customers should bear the capital costs. Prices charged by the water utility should cover capital costs.

In respect of the monitoring process, we recommend that the present system of separating economic regulation from environmental regulation be continued; but a mechanism should be built into the system in order to reconcile the problem of the trade-off between water quality and water charges. Another monitoring mechanism to ensure knowledge of customer preferences is to involve customers in determining pricing and levels of service. The roles of the existing Customer Liaison Group established by the Water Supplies Department in 1993 should be expanded to cover matters concerning water charges and quality expectations. Through the process of customer consultation, the regulator would act as an agent of the consumers to monitor the price and quality of water services provided by the privatized water utility. Formal communication channels between the economic and environmental regulators must be established. In addition, the regulatory responsibilities of the two regulators should be clearly defined, and governance costs should be kept minimal.

Privatizing the Water Supply Industry

Injecting Financial Resources and Improving Productivity

Before handing the industry over to the private sector, the government should consider taking several preparatory actions. One action is to inject sufficient funds into the industry. If the future capital requirements have been met earlier on, then future price increases due to increased investment would be reduced. Alternatively, the regulator should release the business plans of bidding companies to the pubic before selling to the private sector. Each prospective private water utility should submit a long-term investment plan so that the public can be better informed about the expansion of the company and the implications for prices and costs.

Before privatization, another action the government should take is to find ways to enhance the productivity of the Water Supplies Department now. One immediate consequence of such moves could be a mass of redundancies, especially workers at the lower-ranking level. The government must design an appropriate redeployment scheme and compensation package before handing over the water industry to the private sector. Why is this early action recommended? If productivity enhancement has already taken place before privatization, the water monopolist would have less opportunity to raise returns by further cuts in manpower after privatization. This would allow for more accurate projections of future prices and profits, and thus avoid consumer dissatisfaction over excessive profits earned by the water monopolists. Furthermore, if people have seen rising investment, falling prices and improving productivity before privatization, they would have a more favourable attitude towards privatization of the water utility.

Changing the Water Charging System

We affirm the principle that charges should be based on consumption and production costs. The government should consider rebalancing the existing tariff structures: water charges for all users, particularly for non-domestic users, should be increased, and rates paid by property owners and users should be reduced accordingly.

To encourage water conservation, the government may continue to maintain its free allowance provided to small users.

Controlling Monopolistic Behaviour

After privatization, the water supply will be under the control of a private monopolist. Some form of price control may be required to prevent monopolistic behaviour and to enhance improvement in productivity. The purposes of a *price-cap regulation* are to encourage the privatized utility to earn a reasonable return on new investment (the K factor) and to reward efficient production (the X factor). The price cap can contain an automatic adjustment mechanism that reflects cost changes beyond the utility's control. The Hong Kong government can continue to negotiate the terms of the water supply contract with the Chinese authorities. The annual adjustment in the price of China's East River water can be included in the price-cap formula (the Y factor). Hence the price-cap formula can take the following form:

CPI + K − X + Y,

where

CPI = percentage change in consumer price index;
K   = a factor that reflects the percentage change in price required to finance expansion;
X   = a productivity factor (as a percentage):
Y   = cost pass-through.

Allowing Business Diversification and Competition

In order to enhance the profit prospects and to attract private interest in the water supply business, the government may consider allowing the privatized water utility to diversify. As living standards in Hong Kong improve, there is greater demand for value-added water products such as distilled water and mineral water. The water utility can enter into this expanding market and compete with other water suppliers.

Another business opportunity lies in co-operating with China to construct reservoirs in China. Constructing new reservoirs in

China can benefit both Hong Kong and China. While China can provide the land and labour needed in the construction of reservoirs, Hong Kong can provide the engineering support and management skills. This will efficiently utilize the resources of both partners.

As Hong Kong has been relying on China for water supply, some of the small reservoirs in Hong Kong may no longer be very useful in providing water. These reservoirs, together with the associated catchment areas, can be redeveloped for other purposes.

Relying on water from the East River has made Hong Kong subject to monopoly pricing. In the future, the privatized water utility should be encouraged to seek new sources of water in China. The government can modify the cost pass-through factor in the price-cap formula to induce the water utility to seek cheaper sources of water. Hence, a privatized water industry will facilitate the development of a competitive market for water in mainland China.

Pricing the Water Utility

To price the existing assets of the water industry, the concept of "indicative value" may be applied. Regulators in the United Kingdom have used the concept to value existing assets as if the current regulatory regime had not changed. This is considered appropriate because it implies that existing owners would neither gain nor lose from the change in regulatory regime. We recommend calculating the water utility's financial value to the Hong Kong government as if the government were to continue holding the assets and earning the same amount of cash flow (as it would if the department would not be privatized). The "indicative value" of the waterworks assets has been found in preliminary studies to be much lower than the book values.

Employee Ownership and Consumer Acceptance

The government may retain part ownership of the company during the initial stages of privatization. Such a connection between the government and the company would ensure a smooth transition of the public utility into the private sector. A priority programme of share subscription for the employees of the company may be included as part of the privatization process. Share subscription would compensate the employees for the transformation of working status, and induce them to work efficiently because they would see that they are sharing part of the company's profit. Monetary reward to the workers is thus tied to the long-term performance of the company.

If the privatized water utility is to be allowed to invest in new assets to improve water services, then users have to be prepared to face higher charge increases. This is unavoidable if the "users pay" principle is adopted and if the water company is to be run on commercial principles. Price-cap regulation and the take-over threats in the stock market would force management to improve productivity. The net result is that customers would enjoy benefits resulting from productivity enhancement, and taxpayers will no longer be required to subsidize water services. Altogether, we stand to gain from the privatization of the water utility.

Contracting Out the Sewerage System

The experience of contracting out the management of the government tunnels can be applied to the privatization of sewage operations. It is not a good idea to contract out sewage operations to private companies all at once. Just as it did in the case of the government tunnels, the government should first designate a pilot project for private management. After observing the performance of the private operator and assessing the associated problems, the government can then formulate the complete privatization of sewage operations.

It is recommended that the government contract out the various sewage operations to more than one private operator. However, the number of contractors should not be too large in order to allow operators to take advantage of economies of scale in the provision of services. For example, the operation of the High Priority Programme and Strategic Sewage Disposal Scheme Stage 2 should be allocated to one party, Stage 3 and Stage 4 to another party, and the other six treatment plants to still other parties. The assignment of different projects to various parties has several advantages and is important for the success of the privatization programme. The quality of services provided by different operators can be compared and the relative performance level can be assessed. Underperformers will face the threat of replacement by a competitor in the next contract.

The Drainage Services Department should continue to bear the responsibility for planning and constructing the drainage systems and other drainage-related activities such as flood control. The Sewage Service Trading Fund may still be in existence after sewage operations are privatized. There should be no change in the policy of allowing the fund to earn zero return on the department's physical assets. The task of the fund is to collect sewage charges from polluters, and to channel the charges into the payment of the management fee to the private operators. The charges for sewage services will be tied to the bids submitted by the successful bidders, and it is expected that the efficiency gained from actual and potential competition among private operators will help to reduce the pressure for charge increases, and thus lessening the burden on consumers.

Appendix

Table 3.1
Revenue and Cost Structures of the Water Supplies Department
(HK$ million)

	1991–92	1992–93	1993–94	1994–95	1995–96
Revenue					
Chargeable supplies	1,808.4	1,941.9	2,193.6	2,267.9	2,374.8
Contribution from rates	1,281.1	1,616.3	1,659.5	1,955.6	2,127.1
Contribution from Government for free allowance to domestic consumers	287.7	267.6	304.2	347.6	384.3
Supplies to Government establishments	97.0	107.6	126.7	140.5	164.5
Fees, licenses and reimbursable work	40.3	37.2	44.3	45.1	41.3
Interest from deposits	27.5	19.4	18.9	29.2	45.9
Subtotal	3,542.0	3,990.0	4,347.2	4,785.9	5,137.9
Expenditure					
Staff costs	790.0	877.5	1,012.9	1,117.7	1,275.0
Operating and administration expenses	747.4	747.1	831.9	978.3	1,177.1
Bulk purchase of water from China	1,133.9	1,079.1	1,200.4	1,302.3	1,541.1
Depreciation	45.1	279.4	300.4	335.0	400.5
Interest on Government loan	254.8	131.7	134.5	134.3	133.9
Subtotal	2,971.2	3,114.8	3,480.1	3,867.6	4,527.6
Operating Surplus after taxation	476.6	722.0	812.4	860.0	609.7

Source: Treasury, *Operating Accounts of Hongkong Government Utilities, 1992–96.*

Table 3.2
Revenue of the Water Supplies Department, 1980–96
(HK$ million)

Year	Total Revenue	Water Charges	From Rates	Free Allowance
1980–81	752.1	488.3	256.1	
1981–82	832.1	490.1	329.4	
1982–83	1,032.1	582.1	417.9	
1983–84	1,256.2	749.8	475.0	
1984–85	1,383.8	858.8	511.2	
1985–86	1,546.2	973.9	586.4	
1986–87	1,750.3	965.4	597.5	147.2
1987–88	1,937.9	1,092.7	621.9	174.1
1988–89	2,233.5	1,211.9	765.1	202.6
1989–90	2,567.9	1,448.2	822.3	230.7
1990–91	3,112.5	1,702.3	1,092.0	239.4
1991–92	3,542.0	1,177.4	1,281.1	287.7
1992–93	3,990.0	2,049.5	1,616.3	267.6
1993–94	4,346.2	2,320.3	1,659.5	304.2
1994–95	4,785.9	2,408.4	1,955.6	347.6
1995–96	5,137.9	2,539.3	2,127.1	384.3

Source:　Treasury, *Operating Accounts of Hongkong Government Utilities, 1981–96.*

Table 3.3
Water Prices in Hong Kong, 1980–96
(HK$ per cubic metre)

Year	Average Price	Subsidized Price
1980–81	2.05	1.33
1981–82	2.32	1.37
1982–83	2.70	1.52
1983–84	2.75	1.64
1984–85	2.92	1.81
1985–86	3.09	1.95
1986–87	3.19	1.76
1987–88	3.32	1.87
1988–89	3.67	1.99
1989–90	4.02	2.27
1990–91	4.91	2.69
1991–92	5.58	3.00
1992–93	6.33	3.25
1993–94	6.91	3.69
1994–95	7.76	3.90
1995–96	8.69	4.30

Source: Treasury, *Operating Accounts of Hongkong Government Utilities, 1981–96.*

Appendix

Table 3.4
Labour Productivity of the Water Supplies Department, 1979–96

Year	Sales Volume (million cubic metres)	No. of Customers (thousands)	No. of Employees	Labour Productivity	
				(sales per worker)	(customers per worker)
1979–80	335	922	3,694	0.0907	249.59
1980–81	366	994	3,824	0.0957	259.94
1981–82	358	1,053	4,160	0.0861	253.13
1982–83	382	1,167	4,353	0.0878	268.09
1983–84	456	1,259	4,437	0.1028	283.75
1984–85	474	1,307	4,699	0.1009	278.14
1985–86	500	1,400	4,802	0.1041	291.55
1986–87	548	1,483	4,873	0.1125	304.33
1987–88	583	1,576	4,900	0.1190	321.63
1988–89	608	1,685	5,052	0.1203	333.53
1989–90	638	1,773	5,108	0.1249	347.10
1990–91	634	1,849	5,192	0.1221	356.12
1991–92	635	1,909	5,191	0.1223	367.75
1992–93	630	1,946	5,341	0.1180	364.35
1993–94	629	2,023	5,409	0.1163	374.01
1994–95	617	2,075	5,615	0.1099	369.55
1995–96	591	2,120	5,893	0.1003	359.75

Source: Treasury, *Operating Accounts of Hongkong Government Utilities, 1980–96.*

Table 3.5
Labour Productivity of Private Utilities in Hong Kong

A. China Light & Power (CLP)

Year	Sales Volume (megajoules)	No. of Customers (thousands)	No. of Employees	Labour Productivity (sales per worker)	(customers per worker)
1980	29,584	889	6,120	4.83	145.26
1981	30,886	950	7,131	4.33	133.22
1982	33,342	1,011	7,427	4.49	136.12
1983	37,730	1,074	7,234	5.22	145.47
1984	42,231	1,139	7,051	5.99	161.54
1985	45,474	1,200	6,944	6.55	172.81
1986	50,583	1,249	6,698	7.55	186.47
1987	56,409	1,297	6,457	8.74	200.87
1988	60,367	1,347	6,260	9.64	215.18
1989	65,014	1,395	6,329	10.27	220.41
1990	68,666	1,454	6,596	10.41	220.44
1991	77,182	1,509	6,605	11.69	228.46
1992	86,043	1,557	6,587	13.06	236.37
1993	88,108	1,600	6,640	13.27	240.96
1994	81,657	1,655	6,375	12.81	259.61
1995	82,649	1,697	6,378	12.96	266.07

Table 3.5 (continued)
B. Hongkong Electric (HEC)

Year	Sales Volume (megajoules)	No. of Customers (thousands)	No. of Employees	Labour Productivity	
				(sales per worker)	(customers per worker)
1980	10,843	311	2,051	5.29	151.63
1981	11,585	327	2,366	4.90	138.21
1982	12,406	335	2,796	4.44	119.81
1983	14,119	345	2,915	4.84	118.35
1984	14,598	354	2,975	4.91	119.00
1985	15,646	369	2,963	5.28	124.54
1986	17,359	385	2,836	6.12	135.75
1987	19,117	405	2,713	7.05	149.28
1988	20,452	422	2,601	7.86	162.25
1989	21,946	431	2,669	8.22	161.48
1990	23,605	439	2,789	8.46	157.05
1991	24,977	446	2,811	8.89	158.66
1992	25,974	456	2,737	9.48	166.61
1993	27,904	472	2,706	10.31	174.43
1994	29,725	483	2,713	10.96	178.03
1995	30,168	493	2,700	11.17	182.59

Table 3.5 (continued)

C. Hong Kong & China Gas (HKCG)

Year	Sales Volume (megajoules)	No. of Customers (thousands)	No. of Employees	Labour Productivity (sales per worker)	(customers per worker)
1980	3,524	172	1,082	3.26	158.91
1981	4,033	210	1,175	3.43	178.82
1982	4,858	252	1,362	3.57	185.24
1983	5,924	303	1,494	3.97	202.99
1984	6,907	353	1,623	4.26	217.73
1985	7,979	407	1,765	4.52	230.44
1986	9,043	477	1,906	4.74	250.54
1987	10,584	544	2,000	5.29	271.79
1988	12,247	617	2,047	5.98	301.39
1989	13,671	689	2,153	6.35	321.09
1990	15,056	759	2,208	6.82	343.82
1991	16,238	829	2,259	7.19	367.11
1992	18,207	899	2,315	7.86	388.53
1993	19,198	970	2,382	8.06	407.33
1994	20,727	1,041	2,446	8.47	425.43
1995	21,972	1,107	2,476	8.87	446.89

Table 3.5 (continued)

D. Hongkong Telecom (HKT)

Year	Exchange Lines (thousands)	No. of Employees	Labour Productivity (lines per worker)
1983–84	1,573	15,371	102.34
1984–85	1,664	15,361	108.33
1985–86	1,764	15,793	111.70
1986–87	1,877	16,201	115.86
1987–88	2,021	16,755	120.62
1988–89	2,191	17,261	126.93
1989–90	2,345	17,800	131.74
1990–91	2,475	16,279	152.04
1991–92	2,642	15,449	171.01
1992–93	2,820	15,888	177.49
1993–94	2,992	16,039	186.55
1994–95	3,149	16,054	196.15
1995–96	3,275	15,022	218.01

Sources: Annual reports of utility companies; *Hong Kong Energy Statistics*

Table 3.6
Capital Expenditure of Public Utilities in Hong Kong, 1986–96
(HK$ million)

A. The Water Supplies Department

Year	Capital Expenditure Contracted for	Capital Expenditure Authorized but not yet Contracted for
1986–87	615.6	1,044.8
1987–88	529.6	1,180.3
1988–89	597.7	1,283.4
1989–90	917.8	1,187.3
1990–91	593.7	1,344.6
1991–92	442.9	3,824.5
1992–93	1,641.7	2,289.1
1993–94	2,009.5	5,098.0
1994–95	1,760.1	3,093.9
1995–96	1,115.3	3,274.8

B. Electricity and Gas Companies

Year	Capital Expenditure		
	CLP	HEC	HKCG
1986	3,470	1,764	610
1987	2,721	1,106	296
1988	2,164	1,798	328
1989	2,760	2,433	456
1990	3,925	2,439	1,050
1991	6,100	2,667	954
1992	3,969	2,970	816
1993	6,477	3,486	693
1994	7,408	4,668	774
1995	8,715	5,459	786

Table 3.6 (continued)

C. Hongkong Telecom

Year	Capital Expenditure
1986–87	1,280
1987–88	1,576
1988–89	2,387
1989–90	2,676
1990–91	3,018
1991–92	2,204
1992–93	3,620
1993–94	3,532
1994–95	4,045
1995–96	4,331

Sources: Treasury, *Operating Accounts of Hongkong Government Utilities, 1987–96*;
          annual reports of utility companies
Note: See Table 3.5 for company names.

Table 3.7
Labour Cost of the Water Supplies Department, 1980–96

Year	Staff Costs (HK$ million)	No. of Employees	Unit Staff Cost (HK$)	Annual Rise (%)
1980–81	144.3	3,824	39,063.35	
1981–82	175.5	4,160	45,894.35	17.49
1982–83	203.6	4,353	48,942.31	6.64
1983–84	227.5	4,437	52,262.81	6.78
1984–85	275.7	4,699	62,136.58	18.89
1985–86	318.7	4,802	67,822.94	9.15
1986–87	361.1	4,873	75,197.83	10.87
1987–88	445.1	4,900	91,340.04	21.47
1988–89	486.0	5,052	99,183.67	8.59
1989–90	551.9	5,108	109,243.90	10.14
1990–91	689.3	5,192	134,945.20	23.53
1991–92	790.0	5,191	152,157.20	12.75
1992–93	877.5	5,341	169,042.60	11.10
1993–94	1,012.9	5,409	189,646.10	12.19
1994–95	1,117.7	5,615	206,637.10	8.96
1995–96	1,275.0	5,893	227,070.30	9.89

Source: Treasury, *Operating Accounts of Hongkong Government Utilities, 1981–96.*

Table 3.8

Nominal Index of Payroll per Person in Utility Industries, 1990–96

A. The Water Supplies Department

Year	Payroll Index
1990–91	100.0
1991–92	112.8
1992–93	125.3
1993–94	140.5
1994–95	153.1
1995–96	168.3

B. Electricity and Gas Industries

Year	Payroll Index
March 91	100.0
March 92	107.1
March 93	118.9
March 94	132.5
March 95	145.4
March 96	152.5

Source: Treasury, *Operating Accounts of Hongkong Government Utilities, 1991–96*; Census and Statistics Department

Table 3.9
Water Purchase Price

Year	Purchase Cost (HK$ million)	Water Purchase (million cubic metres)	Purchase Price (HK$ per cub. m.)	Annual Rise (%)
1980–81	85.9	172	0.50	
1981–82	122.6	204	0.60	20.34
1982–83	196.2	240	0.82	36.03
1983–84	247.6	255	0.97	18.77
1984–85	222.4	290	0.77	−21.02
1985–86	275.0	325	0.85	10.33
1986–87	314.2	395	0.80	−5.99
1987–88	431.8	417	1.04	30.18
1988–89	681.4	592	1.15	11.16
1989–90	756.7	619	1.22	6.21
1990–91	801.6	604	1.33	8.56
1991–92	1,133.9	724	1.57	18.01
1992–93	1,079.1	662	1.63	4.08
1993–94	1,200.4	668	1.80	10.24
1994–95	1,302.3	650	2.00	11.49
1995–96	1,541.1	692	2.23	11.15

Source: Treasury, *Operating Accounts of Hongkong Government Utilities, 1981–96.*

Appendix

Table 3.10
Returns of Public Utilities in Hong Kong, 1979–96

A. Water Supplies Department

Year	ANFA (HK$ m)	After-tax Surplus (HK$ m)	Target Return (%)	Actual Return (%)
1979–80	3,364.2	262.5	7.0	7.8
1980–81	3,596.4	257.9	7.0	7.2
1981–82	4,075.4	−13.0	7.0	−0.3
1982–83	4,787.1	218.2	7.0	4.6
1983–84	5,631.6	392.7	7.0	7.0
1984–85	6,386.8	444.7	7.0	7.0
1985–86	6,965.1	411.5	7.0	5.9
1986–87	7,581.7	466.5	7.0	6.2
1987–88	7,892.8	394.4	7.0	5.0
1988–89	8,508.0	353.8	7.0	4.2
1989–90	9,180.4	432.2	7.0	4.7
1990–91	9,943.5	578.7	7.0	5.8
1991–92	10,660.1	476.6	7.0	4.5
1992–93	11,443.6	722.0	7.0	6.3
1993–94	12,602.9	805.8	6.9	6.4
1994–95	14,099.9	860.0	8.0	6.1
1995–96	15,711.0	609.7	6.5	3.9

Table 3.10 (continued)

B. Other Private Utilities

(%)

Year	CLP		HEC		HKCG	
	Asset Return	Equity Return	Asset Return	Equity Return	Asset Return (%)	Equity Return (%)
1980	11.68	19.00	12.76	n.a.	14.13	14.72
1981	10.56	19.67	14.11	n.a.	10.13	10.07
1982	9.89	21.02	10.68	n.a.	12.13	14.90
1983	9.49	22.18	10.81	n.a.	15.51	19.13
1984	9.55	24.27	11.10	23.41	18.02	22.68
1985	9.82	26.93	11.62	25.07	16.49	16.91
1986	10.17	25.37	11.38	26.14	14.99	18.40
1987	10.76	25.02	11.36	26.68	17.67	22.40
1988	11.22	24.03	13.07	30.20	21.44	25.73
1989	14.35	23.35	12.56	29.58	22.72	27.61
1990	14.65	23.57	12.41	21.07	21.48	28.30
1991	12.26	24.00	12.43	22.11	20.52	28.76
1992	12.37	23.94	12.49	23.02	21.52	29.22
1993	12.46	24.46	12.58	23.95	15.94	12.15
1994	12.57	25.74	12.41	24.88	13.20	13.37
1995	12.26	26.48	12.03	25.76	13.36	14.79

Source: Treasury, *Operating Accounts of Hongkong Government Utilities, 1980–96.*; annual reports of utility companies

Note: See Table 3.5 for company names

Appendix

Table 3.11
Water Supply and Sales in Hong Kong, 1981–96

A. Water Supply
(million cubic metres)

Year	Total Supply (1)	Water Purchase (2)	From Reservoirs (3)
1980–81	516	172	344
1981–82	493	204	289
1982–83	541	240	301
1983–84	605	255	350
1984–85	628	290	338
1985–86	649	325	324
1986–87	716	395	321
1987–88	762	417	345
1988–89	818	592	226
1989–90	856	619	237
1990–91	880	604	276
1991–92	882	724	158
1992–93	898	662	236
1993–94	916	668	248
1994–95	931	650	281
1995–96	913	692	221

Table 3.11 (continued)

B. Water Sales / Accounted for
(million cubic metres)

Year	Domestic (4)	Trade (5)	Government (6)	Others (7)	Total Sales (8)	Sales / Supply (%)
1980–81	173	140	22	29	366	70.93
1981–82	169	138	24	26	358	72.62
1982–83	177	151	25	27	382	70.61
1983–84	193	186	27	45	456	75.37
1984–85	193	203	32	45	474	75.48
1985–86	200	220	33	48	500	77.04
1986–87	211	251	32	54	548	76.53
1987–88	216	274	36	57	583	76.51
1988–89	219	290	36	63	608	74.32
1989–90	227	316	31	64	638	74.53
1990–91	227	309	31	67	634	72.05
1991–92	230	304	27	74	635	72.00
1992–93	235	295	28	72	630	70.16
1993–94	249	276	30	74	629	68.87
1994–95	255	253	30	79	617	66.27
1995–96	258	227	33	73	591	64.73

Source: Treasury, *Operating Accounts of Hongkong Government Utilities, 1981–96.*
Note: Relations among columns are: (1) = (2) + (3); (8) = sum of (4) to (7); Sales/Supply = (8)/(1)

Appendix

Table 3.12
Reported Incidents of Waterpipe Bursts, 1990–96

Year	Number of Bursts
1990–91	1,056
1991–92	1,141
1992–93	1,202
1993–94	1,108
1994–95	993
1995–96	1,313

Source: *Sing Tao Jih Pao*, 21 September 1996

Table 3.13
Operating Expenses of Lok On Pai Desalter, 1980–91
(HK$ million)

Year	Operating Expenses
1980–81	1.7
1981–82	223.3
1982–83	46.5
1983–84	4.0
1984–85	57.7
1985–86	55.7
1986–87	49.0
1987–88	43.0
1988–89	47.0
1989–90	23.2
1990–91	10.3

Source: Treasury, *Operating Accounts of Hongkong Government Utilities, 1981–91.*

Table 3.14
Water Purchase Price, 1997–2000

Year	Purchase Cost (HK$ million)	Water Purchase (in million cubic metres)	Purchase Price (HK$ per cub. m.)	Increase (%)
1996	1,731.60	720	2.405	n.a.
1997	1,959.75	750	2.613	8.65
1998	2,214.42	780	2.839	8.65
1999	2,498.85	810	3.085	8.67
2000	n.a.	840	n.a.	n.a.

Source: *Express News*, 15 January 1997

Bibliography

1. Armstrong, Mark, S. Cowan, and J. Vickers (1994). *Regulatory Reform: Economic Analysis and British Experience*. Cambridge: The MIT Press, pp. 323–354.

2. Beecher, Janice A. (1995). "The Water Utility Industry," Lecture notes for the Annual Regulatory Studies Programme, held at Michigan State University.

3. Bhattacharyya, Arunava, Elliott Parker, and Kambiz Raffiee (1994). "An Examination of the Effect of Ownership on the Relative Efficiency of Public and Private Water Utilities," *Land Economics* 70, No. 3, (May): 197–209.

4. Booker, Alan (1992). "The Price of Water," *Institutional Investor* 26, No.11: 17.

5. Byatt, I. C. R. (1991). "UK Office of Water Services: Structure and Policy," *Utilities Policy* (January): 164–171.

6. China Light and Power Company Limited (CLP) (1980–95). *Annual Report*.

7. Clarke, Thomas, and Christos Pitells, eds. (1993). *The Political Economy of Privatization*. London and New York: Routledge.

8. Coase, Ronald (1960). "The Problem of Social Cost," *Journal of Law and Economics* 3 (October): 1–44.

9. Conor, Joyce (1993). "Periodic Review," *Investors Chronicle* (May): 55–56.

10. Cook, Paul, and Colin Kirkpatrick (1995). "Privatization Policy and Performance," in their (eds.) *Privatization Policy and Performance: International Perspectives*. London: Prentice Hall, pp. 3–27.

11. Demsetz, Horold (1968). "Why Regulating Utilities?" *Journal of Law and Economics* 11 (April): 55–65.

12. Domberger, Simon (1986). "Economic Regulation Through Franchise Contracts," *Privatization & Regulation — the UK Experience*, eds. John Kay, Colin Mayer, and David Thompson. Oxford: Clarendon Press.

Government units of Hong Kong are listed by name, e.g. Treasury

13. Drainage Services Department (1996). *A Plan for Flood Mitigation* (July).

14. _____ (1996). *Hong Kong Drainage Services Department: Fact Sheet* (March).

15. _____ (1995–96). *Performance Pledge.*

16. _____ (1995). *Sewage Services Trading Fund Newsletter.*

17. _____ (1995). *Your Contribution Towards Cleaner Water* (December).

18. Environmental Protection Department (1995). "Enforcement Activities and Prosecution Statistics under Water Pollution Control Ordinance," *Environment Hong Kong*, pp. 206–207.

19. Finance Branch (1994). *Review on Government Utilities.* Hong Kong Government (October).

20. Grout, Paul (1994). "Popular Capitalism," *Privatization and Economic Performance*, eds. Matthew Bishop, John Kay, and Colin Mayer. New York: Oxford University Press, pp. 299–312.

21. Grout, Paul (1995). "The Cost of Capital in Regulated Industries," *The Regulatory Challenge*, eds. Matthew Bishop, John Kay, and Colin Mayer. New York: Oxford University Press, pp. 386–407.

22. Hartad, Ronald M., and Michael A. Crew (1996). "Franchise Bidding Without Holdups: Utility Regulation with Efficient Pricing and Choice of Provider," Working Paper, Centre for Research in Regulated Industries, Rutgers University (19 October).

23. Hartley, Keith, and Meg Huby (1986). "Contracting-Out Policy: Theory and Evidence," *Privatization & Regulation — the UK Experience*, eds. John Kay, Colin Mayer, and David Thompson, Oxford: Clarendon Press, pp. 284–296.

24. Helm, Dieter, and Najma Rajah (1994). "Water Regulation: The Periodic Review," *Fiscal Studies* 15, No. 2: 74–94.

25. Hongkong Electric Company Limited (HEC) (1980–95). *Annual Report.*

26. Hong Kong and China Gas Company Limited (HKCG) (1980–95). *Annual Report.*

27. Hong Kong Government (1946–96). *Hong Kong Annual Report*, chapter on public utilities.

28. Hong Kong Telecommunications Limited (HKTel) (1988–95). *Annual Report.*

29. Hunt, Lester C. and Edward L. Lynk (1995). "Privatization and Efficiency in The UK Water Industry: An Empirical Analysis," *Oxford Bulletin of Economics and Statistics* 57, No. 3: 371–388.

30. Jenkins, Graham (1987). "Water Pollution: The Time's Long Past for Dithering," *Hong Kong Business Today* (June): 14–15.

31. Johnson, Anthony (1991). "Sewage Master Plan," *Asian Architect and Contractor* 21, No.11 (November): 25, 42.

32. Kolbe, A. L., J. A. Read, and G. R. Hall (1984). *The Cost of Capital: Estimating the Rate of Return for Public Utilities.* Cambridge, Mass.: The MIT Press.

33. Leung, P. K. (1995). "Privatization of Tunnels — from Management and Cost-Benefit Perspective, with the Experience of the Aberdeen Tunnel," Final Year Project, Department of Business Studies, Hong Kong Polytechnic University.

34. Littlechild, Stephen (1983). *Regulation of British Telecommunications' Profitability.* London: HMSO.

35. Lok, Mimi (1995). "There Is No Free, Clean Water in Hong Kong," *Varsity* (December): 17.

36. Lynk, E. L. (1993). "Privatization, Joint Production and the Comparative Efficiencies of Private and Public Ownership: The UK Water Industry Case," *Fiscal Studies* 14, No. 2: 98–116.

37. McMaster, Robert and John W. Sawkins (1993). "The Water Industry in Scotland — Is Franchising Viable?" *Fiscal Studies* 14, No.4: 1–13.

38. Mayer, Colin and Shirley Meadowcroft (1986). "Selling Public Assets: Techniques and Financial Implications," *Privatization & Regulation — the UK Experience*, eds. John Kay, Colin Mayer, and David Thompson. Oxford: Clarendon Press.

39. Meder, Ross (1993). "Hong Kong–PRC Interdependence: Convergence Comes By Pipeline," *Amcham* 25, No. 4 (April): 23–26.

40. Meredith, Sandra (1992). "Water Privatization: the Dangers and the Benefits," *Long Range Planning* 25, No. 4: 72–81.

41. Sawkins, John W. (1995). "Yardstick Competition in the English and Welsh Water Industry: Fiction or Reality?" *Utilities Policy* 5, No. 1: 27–36.

42. Scott, Sue (1995). "LRMC and Charging the Polluter — The Case of Industrial Waste Water in Ireland," *Utility Policy* 5, No. 2: 147–164.

43. Sewage Services Trading Fund (1995/96). *Annual Report.*

44. Siddall, Linda (1986). "Fragrant Harbour?" *Welfare Digest* (March): 2–3.

45. Taylor, Michael (1991). "Hongkong's Public Monopolies Need a Dose of Salts: Ripe for Sale," *Far Eastern Economic Review* (2 May): 43.

46. The Bureau of National Affairs (1994). *International Environment Reporter*, section on Hong Kong (November): 160, 162–164.

47. The Bureau of National Affairs (1995). *International Environment Reporter*, section on Hong Kong (July): 119–120.

48. Treasury (Hong Kong Government) (1980–96). *Operating Accounts of Hong Kong Government Utilities: Water Authority.*

49. Utt, Ronald D. (1991). "Privatization in the United States," *Privatization and Economic Efficiency — A Comparative Analysis of Developed and Developing Countries.* Hants, U.K.: Edward Elgar, pp. 73–86.

50. Vass, Peter (1993). "Water Privatization and the First Periodic Review," *Financial Accountability and Management* 9, No. 3 (August): 209–224.

51. Vaughan, Lauri (1988). "Why The Fragrant Harbour Stinks," *Executive* (February) : 35–36.

52. Vickers, John and George Yarrow (1988). *Privatization: An Economic Analysis.* Cambridge, Mass.: The MIT Press, pp. 171–194, 387–423,

53. Walder, Gary (1993). "Private Sector Lobbying Pays Off," *Britain in Hong Kong: The Journal of the British Chamber of Commerce*, (November/December): 15.

54. Water Authority (1996). "1996–97 Water Tariff Revision Proposal," Discussion Paper for the Legco Panel on Planning, Lands and Works (19 November).

55. Water Services Association (UK), and Water Companies Association (1991). *The Cost of Capital in the Water Industry*, Vol. 1–3. London: Water Services Association.

56. Water Supplies Department (1993). *Hong Kong's Water* (October).

57. Water Supplies Department (1996). *Hong Kong Water Supplies Department* (January).

58. Water Supplies Department (1996). *Principal Functions and Services of Water Supplies Department* (January).

59. Water Supplies Department (1996). *Water Charges Calculations for Domestic Supplies* (April).

60. _____ (1996). *Water Supplies in Hong Kong* (March).

61. _____ (1996). *Water Treatment and Quality Control in Hong Kong* (January).

62. Williams, Martin (1992). "Pricing in The Case of Privately-Owned Water Utilities," *Journal of Applied Business Research* 8, No. 4: 97–105.

63. Williamson, Oliver E. (1976). "Franchise Bidding for Natural Monopolies — in General and with Respect to CATV," *Bell Journal of Economics* 7 (Spring): 73–104.

64. Winnifrith, Tom (1994). "The Algebra of H$_2$O," *Investor Chronicle* (8 July): 12.

65. Wright, Mike, Steve Thompson and Ken Robbie (1994). "Management Buy-outs and Privatization," in *Privatization and Economic Performance*, eds. Matthew Bishop, John Kay, and Colin Mayer. Oxford University Press, pp. 313–335,

Index

About the Authors

Pun-Lee Lam, PhD (Bristol), is Assistant Professor at the Hong Kong Polytechnic University. His research interest lies in the area of government regulation of public utilities, and his PhD dissertation was on the government control of electricity companies. Some of the results have been published in a book in this Series, *Competition in Energy*. The results of his research on government policies have been published in various international journals that address policy issues.

Yue-Cheong Chan, MPhil (The Chinese University of Hong Kong), is Assistant Professor at the Hong Kong Polytechnic University. He is interested mainly in doing empirical research on asset pricing models, volatility behaviour, and information dissemination processes in financial markets. The results of his research have been published in refereed academic journals, conference proceedings, magazines and leading newspapers.

The Hong Kong Economic Policy Studies Series